Maupassant's Fiction and the Darwinian View of Life

Currents in Comparative Romance Languages and Literatures

Tamara Alvarez-Detrell and Michael G. Paulson
General Editors

Vol. 143

PETER LANG
New York • Washington, D.C./Baltimore • Bern
Frankfurt am Main • Berlin • Brussels • Vienna • Oxford

Laurence A. Gregorio

Maupassant's Fiction and the Darwinian View of Life

PETER LANG
New York • Washington, D.C./Baltimore • Bern
Frankfurt am Main • Berlin • Brussels • Vienna • Oxford

Library of Congress Cataloging-in-Publication Data

Gregorio, Laurence A.
Maupassant's fiction and the Darwinian view of life / Laurence A. Gregorio.
p. cm. — (Currents in comparative Romance languages and literatures; v. 143)
Includes bibliographical references and index.
1. Maupassant, Guy de, 1850–1893—Criticism and interpretation. 2. Darwin, Charles, 1809–1882—Influence. 3. Evolution (Biology) in literature. 4. Naturalism in literature. 5. Evolution in literature. I. Title. II. Series.
PQ2357.G74 843'.8—dc22 2004027483
ISBN 0-8204-7637-4
ISSN 0893-5963

Bibliographic information published by **Die Deutsche Bibliothek**.
Die Deutsche Bibliothek lists this publication in the "Deutsche Nationalbibliografie"; detailed bibliographic data is available on the Internet at http://dnb.ddb.de/.

The paper in this book meets the guidelines for permanence and durability of the Committee on Production Guidelines for Book Longevity of the Council of Library Resources.

275 Seventh Avenue, 28th Floor, New York, NY 10001
www.peterlangusa.com

Printed in Germany

This book is dedicated with gratitude to great mentors and great friends,

Professor Thomas J. Donahue, Saint Joseph's University, Philadelphia

and

Professor Jean V. Alter, Emeritus, University of Pennsylvania.

Contents

Acknowledgments

I wish to express my gratitude to Gettysburg College for the sabbatical and the material support necessary to make this book a reality. I am also grateful to the editors and staff of Peter Lang Publishing for their help throughout the publishing process.

To my colleagues in the Biology Department of Gettysburg College, Ralph Sorensen and Steven James, I offer my sincere thanks for their unflagging patience in answering my many questions and for their great moral support during the drafting of the manuscript. My thanks likewise to my colleagues in my own Department of French and Italian at Gettysburg, especially Robert Viti, Florence Ramond Jurney and Elizabeth Richardson Viti, for their enthusiastic support and for their advice and help with questions of literary history. I gratefully acknowledge the technical assistance of Sue Welsh, Gina Rouiller and Kim Breighner in the solution of innumerable formatting problems during the preparation of the manuscript. My appreciation also goes to Susan Roach and the Interlibrary Loan staff at Gettysburg College for indispensable help during my research. To a career-long teacher, friend and sounding board at Penn, Gerald Prince, I also tender my thanks.

Finally I acknowledge with love and immeasurable gratitude the constant support and inspiration given by my wife Marcia and our children, Joseph, Julie and Elena. Their interest and encouragement have made the entire project all the more enjoyable for me. Many, many thanks to them.

L.A.G.

Introduction

Let me state at the outset that I am not a biologist. I don't even play one on TV. My professional field is French literature, so how is it that the following study, which betokens a keen interest in Darwin and evolutionary theory, could come about?

This is one of the glories of teaching in a liberal arts institution. Some twenty years ago, I had the good fortune to be teaching a section of a multi-disciplinary, general course required at the time of all first year students. It was a course on the general topic of "Revolution," intended to expand the horizons of students' intellectual inquiry as well as the limits of their intellectual baggage, but it ended up having that effect on me. Part of the course's syllabus dealt with the revolutionary idea of Darwin. Here I first encountered the marvelous writing of the remarkable scholar and popularizer (in the very best sense) of evolutionary thought, Stephen Jay Gould. From this first encounter, I was hooked. In the intervening years, I have had the hobby of reading everything pertinent to evolutionary biology that I could lay my hands on—everything, that is, accessible to the non-biologist. Fortunately, there is a world of fine reading available. Authors like Gould, Richard Dawkins, Cynthia Mills, Steve Jones, Jared Diamond, Jonathan Marks, Niles Eldredge, and Jonathan Weiner, to name only a few, have made a fascinating ideological landscape approachable and understandable to a general readership.

What could be so important about the idea of evolution? And why should it be of general interest? I asked myself the very same questions on an occasion that remains vivid in my memory of a time more than 30 years ago. I was a college freshman myself, and it was the last day of class in a general biology course that I was taking. In concluding his final lecture, the professor said something that stayed with me. His words were, "If I can leave you with one thing to remember long after this course is done, it is this: evolution happened." It struck me as odd, especially since the course had dealt with the usual botany and anatomy of general biology courses, and not with the issue of evolution per se at all. So why would a last lecture conclude that way?

This college freshman was still naïve enough not to know what a truly great idea evolution is, and what a flashpoint it has always been for those who don't want to recognize it. Biologist Theodosius Dobzhansky published an article in 1973 whose title states the case succinctly, "Nothing in Biology Makes Sense Except in the Light of Evolution." Aside from the fact that Darwin's idea has shaped the study of biology at the level of organism and species as well as in microbiology, it is the only credible conceptual framework for the scientific study of life. It has had tremendous influence on thought outside the immediate realm of biology, in philosophy, geology, anthropology, psychology, linguistics, economics and social policy, religion, politics, and in literature.

Something so influential should be clearly understood, yet Darwinism has been the target of attack and disinformation since the *Origin of Species* first appeared in 1859, and always for reasons having nothing to do with science. Objection has almost never come from scientific or philosophical imperative, but from an agenda. Simply put, the "creationism" that begs for attention for all the fuss it "creates," brings literally nothing to science or to philosophy—no data, no experimentation, no validation, no support from the fossil record, no support from DNA typing, from embryology or from comparative anatomy. I hope that the reader of this book will come to appreciate that "creationism" would not bring anything to the reading of Maupassant, either.

Darwinism, on the other hand, does. The Darwinian reading in the pages to come takes account of the fact that evolutionary thought had already begun to capture the imagination of thinkers and writers by the time Maupassant arrived on the literary scene. It also takes account of the fact that, to paraphrase Dobzhansky's dictum, some things in Maupassant make sense only in the light of evolution.

About a year prior to starting this project, I taught another first-year course, this one a seminar designed by me, entitled "The Evolution Revolution," a survey of the influence of Darwin across the liberal arts. While crafting the syllabus for this course, I was anxious to include a component on my own field of scholarly interest. As I considered a number of choices for a literary text which was suitable for length, interest and thematics, I settled on Guy de Maupassant's "La Ficelle" (more precisely, on its translated version, available in a Penguin edition of *Selected Short Stories*). While preparing the class meeting on "The Piece of String," and seeing my students' growing interest in the seemingly odd notion of applying the model of Darwinian thought to the reading of a literary text, the idea slowly grew on me to do a fuller study of Maupassant whose writing seemed to lend itself well to this interpretive model.

Maupassant's complete works of fiction have been gathered into a definitive edition that is marvelous not only for its establishment of the texts, but also for the editorial notes, chronologies and critical apparatus that accompany the texts. I speak of the Gallimard *Pléiade* edition in three volumes done by Louis Forestier. All references to Maupassant's fiction will be to this edition and will be noted parenthetically in the pages to follow, according to these abbreviations: "R" will stand for the volume entitled "Romans," "CN1" will denote "Contes et nouvelles, tome 1," and "CN2" will refer to "Contes et nouvelles, tome 2." Thus, the notation (R, 332) will refer to page 332 of the volume "Romans," (CN2, 995) will refer to page 995 of the second volume of "Contes et nouvelles," and so on.

Naturalism presented itself as the literature of scientism, the fictional laboratory of life that was to portray the world as it is, in a literary setting. No science was more in the forefront or generated more public interest in the 1880's than Charles Darwin's great new idea of evolutionary study, so no science had more influence on the outlook of the Naturalists than that of the *Origin of Species* and the *Descent of Man*. If it may be fairly said that Maupassant's bond with the Naturalist school weakened over time, it must also be recognized that his novelistic vision remained that of the dispassionate, objective observer influenced by the determinism and pessimism of Schopenhauer and the ideological materialism of Spencer and others: in short, he maintained his own variety of scientism in the creation of his novelistic world. In that world, the uncompromising and amoral realities of nature—natural selection, the survival of the fittest, heredity, the inhospitableness of the environment, the relatedness of species and the animal side of humanity—are facts of life. In ways which we shall study, Maupassant's view of that life was very much the same as Darwin's.

Scholarship on the fiction of Maupassant has long been very active, and lately it has continued apace on several thematic fronts. Most recently, the topic of gender (feminine and masculine) in Maupassant has been of interest (see P. Hadlock, 2003 and 2002, M. Donaldson-Evans, 2002, C. Stivale and S. Johnson and M. Berta). Likewise issues of irony (H. Färnlöf), self and other (S. Mamoon), and taboo (F. Place-Verghnes) are among those which have been profitably studied in the year or so prior to this writing. With this book I propose a new reading of Maupassant's fiction, one that by no means preempts or discounts readings in other avenues of inquiry. Instead, this reading proposes to read Maupassant through the lens of evolutionary theory, to bring to light aspects of this Naturalist's novel vision that have not yet come to critical attention. Darwin, at the end of *Origin of Species*, said of his own theory, "There is grandeur in this view of life," and I propose in this book to study such remarkable things as they figure in Maupassant's novels and short stories.

Chapter 1:

Darwinism and Its Evolution

To begin a chapter on what Darwinism is, and particularly on what it also is not, I am reminded of the opening of astronomer Carl Sagan's last book, *Billions and Billions*. In his usual affable way, Sagan began by asserting that he "never said it." He was referring to the catch-phrase that had come to identify him in the popular psyche, entirely through the caricatured impersonation of a late-night TV comedian who imitated Sagan's voice intoning the words "billions and billions" repeatedly. But Sagan had never, to his memory, used the phrase, and still people linked it to him.

Everyone knows who Darwin was and everyone has an idea of what his theory set forth. The principles of evolutionary theory, if not the details, are very much a matter of common knowledge. Nonetheless there are some things widely attributed to Darwin, things which he might point out he had "never said." An interesting one is the term "evolution" itself. Coined in common parlance surrounding his groundbreaking work, the word never actually appears in his 1859 manifesto, *The Origin of Species*. The closest he comes to using the term there is the book's very last word, the participial form "evolved." To the word *evolution* itself, Darwin seems to have preferred terms like "descent with modification," "Divergence of Character" or "adaptation" in his early writing on the topic.[1]

A more significant misattribution is that of the phrase "survival of the fittest." The concept was certainly akin to Darwin's thought, but the turn of phrase seems to belong to Herbert Spencer and developed more in the context of what became known as Social Darwinism than in the sphere of biology itself. Darwin never objected to the phrase and evidently found it succinct and accurate enough, adding it to later editions of the *Origin*, but in his own earlier writing on evolution he denoted the concept as the "Struggle for Life" resulting in the extinction of "less-improved forms." In later writing (e.g. chapter 5 of his *Descent of Man*) the actual phrase "survival of the fittest" is part of

Darwin's vocabulary. Spencer's concern, at any rate and as we shall later have occasion to note, was more with sociological dynamics than with principles of biology and differential rates of survival in nature.

Darwin is the name associated of course with the notion of evolution, understood in its general biological sense as the theory that species of organisms change or adapt morphologically over time to be different from their ancestors. But there are two historical points of interest to be raised here. First, the concept of the mutability of species was by no means new with Charles Darwin. In fact, his own grandfather, the physician and amateur naturalist Erasmus Darwin, had devoted much energy to promoting the theory that species changed over time. What was revolutionary in his grandson's teaching was his proposal of the *mechanism* of evolutionary change, natural selection, and the conclusion that all extant and extinct forms of life are historically related.

The other point of note is that Darwin, while slowly arriving at his overall theory over years of reflecting on his observations of data, and knowing well the implications of his thought for philosophy and religion, kept mum about his findings except to a few friends. Out of the blue in 1858 he received a letter from one Alfred Russel Wallace (1823–1913) who had independently arrived, to Darwin's astonishment, at the very same conclusions. At his friends' insistence, and fearful that he would be scooped, Darwin rushed his *Origin* to press within a year, but he was considerate enough to make the public presentation of the theory a paper authored jointly with Wallace.

The *Origin* itself, a mere sketch by Darwin's standards, is widely assumed—logically, given the work's title—to deal with the question of the origins of life. In fact Darwin's work does not address this topic at all, possibly because he wished to circumvent the firestorm pursuant to a direct challenge to religious belief in the creation, and certainly because the issue would be more speculative than inferential, for lack of scientific evidence. Instead, Darwin's writing treats the adaptation, change and interrelatedness of species beyond the most general of speculations with respect to the origins of life itself.

While Darwin's work is necessarily materialistic in the philosophical sense, it is wrong to assume that his writings are stridently, or even overtly, atheistic. True, in his private correspondence he admitted that he was not a man of religious faith and that religious belief could not trump the scientific in his mind. But he did endeavor to make concessions, as the last paragraph of the *Origin* makes clear: "There is grandeur in this view of life, with its several powers, having been originally breathed by the Creator into a few forms or into one..." (459–460).

Important misconceptions about the substance of Darwin's theory involve inferences wrongly drawn where teleology and human ontology are concerned. Common opinion has jumped to the conclusion that, since evolution means change over time, it must also mean that change is directed towards something. Perhaps it is due to Victorian optimism with regard to "progress," or perhaps it is in our nature to put a moral face on change, but Darwinian biology makes no provision for the moral worth of a surviving species. Put another way, a surviving species is not somehow "better" morally for having survived; it is only better adapted as Darwin would see it. And thus Darwin would have it that the proper inference to be drawn is that there is no ideal or goal towards which life is evolving. This is an important distinction for the literary application of evolutionary thought, as we shall see, for the Naturalists in general (and Maupassant in particular) put forward a fairly pessimistic view of the road down which humanity is traveling. For Darwin, evolution is blind and amoral, having nothing to do with perfecting life forms in a moral sense.

Hence the grave misconception about human ontology. It is often assumed (and certainly has been since Darwin's Victorian time) that if Darwin is right and evolution has produced humanity as it is, then evolution must not only have been pointed towards us, but it must also have somehow privileged us atop a historical ladder of increasing perfection. Darwin's own thought puts the lie to this (given his refusal of the teleological model for evolution, and given his metaphor of the tree of life as we shall see shortly), but it is nonetheless commonly inferred that Darwinian theory ends with *homo sapiens,* having arrived at consciousness, language and the opposable thumb. The true application of Darwinian theory takes humanity down from its self-appointed pedestal, making of it merely another twig on the tree of species in history. Thus, gone at once are the related notions of a teleologically directed force of evolution and a privileged species of humankind at the upper end of a ladder of increasing worth and perfection. Still, misconceptions sometimes surface in the form of opinions that evolution has reached its chronological end (or even its pinnacle) with the advent of humankind, or that the only really important application of evolutionary theory is in the area of anthropology. When, in chapter 6 of the *Origin,* Darwin addresses the matter of what he calls "organs of extreme perfection" (214–224) he does so to make an argument about the mechanism of evolution, and not to editorialize on some moral gradient of values. He is speaking of functional, not moral, perfection.

There will follow here a synopsis of Darwin's evolutionary theory which will also serve, in part, as a sketch of the topics addressed sequentially in the rest of this book as they relate to the study of Maupassant's fiction; obviously some of the points are less pertinent to a study of Maupassant than others. I also hasten to point out that this is not a comprehensive survey of Darwinism, since there are many topics of technical, biological import (embryology, or the imperfection of the geological record, for example) which have no bearing on a non-technical overview of Darwin or on the study of a Naturalist writer of fiction. These I shall pass over with an eye towards summarizing the tenets of Darwin's thought that are both salient and pertinent to the reading of Maupassant and necessary for a general understanding of Darwin. For the rest, I refer the curious reader to the supplementary readings listed in the bibliography (any of the works of Gould, Dawkins, Cynthia Mills, Berra, Jones, Dennett, Diamond and Marks) or to the unabridged *Origin of Species* itself.

Darwinism

Random mutation and the mutability of species

The first observation which set Darwin's thought in motion was that individuals within a species, while they display obvious uniformity of characteristic, display a degree of variability as well. No members of a species are completely identical. This variability, along with the peculiarities of geographic distribution (see below) and comparisons made between extant forms and fossils, led Darwin to the unavoidable conclusion that historical boundaries between what we recognize as species must be fluid. Varieties of one species which have become geographically isolated from one another are on the path toward becoming separate species. In a word, species do not remain static over time. Morphological changes occur that evidently spread throughout a species over its history.

Darwin's thought on this matter was greatly influenced by his observations in what he came to consider a theater of evolution in progress, the Galapagos Islands. He was struck by the varieties of finches he collected there, each showing anatomical differences peculiar to the particular islands where they lived. He drew the conclusion that there is some force in nature which leads species to change, and to change permanently so that groups previously related eventually become henceforth distinct. This is Darwin's principle of divergence: in what Darwin calls the "economy of nature," the successful and dominant forms in a species tend to adapt to diversified places, diverging from previously similar life forms and creating a branching pattern of relationship.[2]

Variation within a species, as a function of biological heredity, appeared to Darwin to be random in nature, and certainly not dictated by any preordained principle or overseeing power. Likewise, variability appeared to him to be as likely to be favorable as unfavorable to the survival and prosperity of the organism; no moral direction and no order of progress accompany the blind force of change (or changeability) that typifies the development of every living species.

The model of the tree or bush

In searching for a metaphorical illustration of his principle of divergence, Darwin hit upon the model of the "tree." Each species would be represented by a twig or branch on the tree, the outermost ones denoting the extant species, ready to continue the growth pattern outward in time. The historical relationship between species could be traced back through the tree, each species deriving (or branching apart) from a related species, and so forth back through the ages.

The implications are clear and were consciously chosen by Darwin: all life is interrelated, and all forms of life bear common ancestry. No matter how far apart two twigs on the tree may be, no matter how disparate their appearances or modes of life, it is possible, in Darwin's theory, to trace their lineages back to a point of common divergence. This is of course one of the deductions of Darwinism that caused an uproar still resounding in a few quarters outside of science: if in fact all species, living and extinct, are historically related and derived from earlier forms, then the notion of "creation" (as set forth in the Bible, for example) is substantially affected. In addition, Darwin's tree displaces a model of ontological interpretation—the model of the ladder or the Great Chain of Being—which had held sway in Western thought for many centuries.

Biologists today seem to prefer the metaphor of a bush to Darwin's original tree. For them, the trunk of a tree conveys a misleading impression, larger in size than all other branches and playing a central role in the development of the rest of the tree. A bush, on the other hand, gives a better picture of the evolutionary process, stemming back to a central point near the ground and branching out in all directions from there. This is merely a question of clarity in illustration.

But as we discuss the implications of this model of thought, we must not overlook another vital theoretical conclusion. If the history of life is accurately portrayed by the bush or tree model, it is an obvious deduction that all life derives from a primordial organism. For philosophers, this is much more of a hot issue than for biologists, and Darwin—knowing well the heat that it would generate—tried to make a concession. In the last sentence of his *Origin of Species* he writes "There is grandeur in this view of life, with its several powers, having been originally breathed into a few forms or into one;..." (459–460), leaving at least the appearance of an open door to belief in creation of multiple life forms. Needless to say, it was not enough to appease his critics.

One last point concerning the branching pattern of species development: as with the outward growth of a bush, so too are species observed to develop in one direction over time. This is to say that the path of evolution is irreversible. Once a species has achieved separateness, it cannot revert to prior forms (this in keeping with the functional definition of *species*, a group of like organisms capable of interbreeding).

The struggle for life

Darwin's observations of nature led him to think of the history of life as a history of struggle for survival in a world which is blind and indifferent to the needs of life. This indifference, and the struggle for survival that it entails, are cardinal points of evolutionary theory for they are the circumstances which set the process of evolution in motion. They provide the environmental pressure to which life responds by adapting, and thus evolving.

In the Galapagos particularly, Darwin had begun to grapple with the question of life in perpetual struggle for survival. But it was some years after that he was finally able to integrate the concept into a comprehensive theory. After rereading the book of economist Thomas Robert Malthus, *Malthus on Population*, Darwin reasoned that the same mathematical principle proposed in that study could also be applied to life in the wild. In a nutshell, Malthus proposed that populations reproduce geometrically while resources to support them grow only arithmetically—meaning simply that the growth of human population will eventually test the environment's ability to sustain it; yet, populations tend to remain constant, betokening some means of control in population growth. Applying this formula to biology, Darwin concluded that, since organisms tend to reproduce maximally for the sake of propagation, the environment invariably "responds," exercising a check on the growth of population by the simple fact that resources for survival are limited. A struggle thus ensues among species as well as within species, for survival and prosperity (i.e. reproduction, ultimately) in an environment where life can either thrive or die out depending on the species' (or individual's) ability to get along.

The key point here is that resources are limited in a finite world, and with life's tendency to populate indefinitely, competition is the law of nature. Left at this juncture, it is easy to see how moralists would find fault with Darwin: nature so described is utterly amoral. Furthermore, it is a small jump to what has become known as Social Darwinism, and another small jump from there to the abuses to which the eugenicists put biological theory. Darwin had nothing to do with any of these tangents, limiting his attention to biology alone, but willy-nilly they became parts of the Darwinist tradition. Of Social Darwinism, more later.

Natural selection

The central and innovative point of Darwin's theory was the concept of natural selection. Ideas of the mutability of species and extinction, and the geological notion of the ancient earth (whose vast expanses of time would be required for Darwin's theory to work) were by no means new. But it was the mechanism of evolution that was missing until Darwin put adaptation together with the Malthusian principle of competition for survival. Only then did the theory make sense.

Competition for limited resources and successful avoidance of predators, parasites and disease necessarily produce differential rates of survival: some organisms are better suited to life in adversity than others in their species. The adversity of the environment exercises the blind, amoral pressure on life to survive, and that adversity takes on any number of forms—sparsity of food, overpopulation, successful survival of predators, hostile climate, etc. In the face of this, any number of a given species *might* survive, but those with some physical advantage—however slight—are more *likely* to survive. This statistical likelihood was the difference for Darwin, and it explained for him how change or adaptation fit into the picture of the life and evolution of species. The point was that any anatomical or behavioral feature, if it were at all favorable to survival in an essentially hostile environment, would confer some degree of advantage, however slight.

We have already noted that mutation occurs randomly; the study of genetics since Darwin's time has gone far to explain the mechanics of this phenomenon. The central issue of natural selection is this: blind nature brings pressure to bear on life forms, and those life forms will show mutation from generation to generation; some of the changes will be of no help (or worse), and natural environmental pressure will work (or "select") *against* the survival of those features; but some of the changes, by dumb luck, will be of help in the organism's struggle for life, and natural environmental pressure will "select" *for* this type of adaptation. The concept does appear to personify nature, what with the element of "selection" brought into the question, but it is crucial to understand nature as a source of constant pressure on life, hence the utility of the notion of nature selecting for or against a given adaptation. If organisms with a favored trait tend to live longer or better, and those without the trait tend in any measure not to fare so well, the point is clear.

Reproduction and heredity

It is easily understandable that reproduction and heredity are the upshot of the principle of natural selection (which is itself the cardinal point of evolutionary theory). Differential rates of survival logically translate into differential rates of success in reproducing and passing all traits, including new and favored ones, on to a new generation. If natural selection is the mechanism of evolution, reproduction and heredity are the stuff with which it works. Success in reproduction is, in the end, success in evolution. Subsequent generations will inherit favored traits, and a simple mathematical model would illustrate how—given time and differential rates of reproductive success—a favored trait can become the norm, producing a changed or evolved species in a circumscribed environment.

The actual mechanics of heredity were unknown to Darwin, and they eluded him. This is the one area of his theory which he had wrong, as through his career he proposed a couple of erroneous genetic models by way of explaining the heredity of traits. It is one of the more interesting bits of Darwinian lore that the one book he needed to untie that knot for him—Gregor Mendel's study of heredity in peas—remains to this day on his bookshelf at his country house in Down, with the pages still uncut. Had Darwin gotten around to reading the work, he certainly would have understood the implications of particulate genetics, that is, the combining and recombining of genetic particles in more or less predictable numbers.

There is one item to underscore in this regard, and it relates to the history of science in France. Jean-Baptiste Lamarck (1744–1829) was a biologist of considerable influence and a great believer—before Darwin—in the mutability of species. He will always be remembered as having championed the theory of the heredity of acquired traits, that is, the belief that if an individual organism acquires a trait (say, strength in a limb from its own behavioral overuse of that limb), it will somehow pass that acquired trait on to offspring as part of its genetic heritage. Stephen Jay Gould has gone to some pains to absolve Lamarck, showing how the French scientist backed away from the flawed theory late in his career (Gould, 2000, 115–143), but it must be noted here that Lamarck's thought on the heredity of acquired traits was wrongly associated with Darwin during the nineteenth century in French popular culture. Darwin for his part, however, put no stock in the notion and believed only in biochemical heredity.

At any rate, heredity and parentage are key factors in evolutionary thought, and much attention has been accorded them in and around science since Darwin. One notable extrapolation of Darwinian thought has been developed by Richard Dawkins since the 1970's, holding that a vital property of genes is to program the need for their own propagation into the organism bearing them.[3] The stimulus to reproduce thus is envisioned as an instinctual, biochemically driven force related to the species' struggle for life. Interestingly, a theory of altruism has come about in evolutionary biology, viewing it as a means of assuring or facilitating the successful reproduction of the genetic line, and hence the survival of the species. In this line of thought, a not illogical and close deduction from Darwin's writing on instinct and the social insects (chapter 7 of the *Origin*, 234–263), parentage is a function of survival of genetic heritage, and the survival of the species (or more particularly, the genetic line) takes precedence over the survival of the individual, as a matter of unconscious biological impulse. The implications of this thought are of interest outside of biology, too, in consideration of the social dimension of Darwinian theory.

Sexual selection

Darwin evidently sensed that natural selection alone either could not by itself account for all of evolution, or would not satisfy all skeptics as the lone mechanism of evolutionary change. Thus he devoted a large part of his book, *The Descent of man and selection in relation to sex,* [4] to the study of what he termed sexual selection. Simply put, this has to do with the development of physical traits which give sexually reproducing organisms an advantage over rival members of their own species and sex in the drive to reproduce (or to be chosen by the opposite sex for mating). If reproduction is the means by which evolution is effected, then success in being chosen for reproduction must also be a factor for study in evolutionary theory. This is the area of biology which weighs, for example, the advantages of the peacock's cumbersome tail in luring females, against the disadvantages it poses for allowing movement and warding off enemies.

Under this rubric, Darwin also considers the issue of human morality. In chapter 5 of the *Descent* he makes the point that moral behavior would likely be selected *against* since it acts to repress the reproductive urge; but he also makes the point that other moral behavior serves the purposes of altruism and thus contributes to the good and survival of the species community (147). Speaking anthropologically, Darwin asserts that a high standard of morality may give no advantage to the individual, but does confer an "immense advantage to one tribe over another"(150). However he does go immediately to assert that *progress* in society is not the rule historically (150). These points will be of interest to the study of Darwin's influence on social theory and on the literature of the Naturalists.

Concentration and periphery

An element of Darwin's theory that could be seen as having an influence on sociological thought is the question of the geographical concentration of a species. Briefly put, Darwin's observations led him to conclude that high concentration of members of a species in a circumscribed geographical area yields pressure for genetic conformity, resulting in less frequent change or speciation; conversely, less concentrated life on the periphery of that area tends to yield less pressure for genetic uniformity, resulting in more opportunity for species change. Chapters 11 and 12 of the *Origin* deal with topics of geographical distribution and migrations. In general, the focus on the question of the dynamics of population density in science could clearly fuel interest in the same issue in sociological thought and fictional literature.

Human ontology

Lastly we reiterate in some measure the question of humanity's status in light of Darwin's new theory. Inferences can be (and, since Darwin's time, have been) readily drawn regarding where humanity stands in this world view. Humanity is now described as another species of animal, having derived from some earlier non-human species. No longer does it occupy an exalted status on a ladder of ontological worth, as it had done in Western thought prior to Darwin. Gwenhaël Ponnau speaks of

> ...la conception et la perception d'une histoire dont l'homme est non pas, non plus le roi, mais le sujet précaire, *par nature* éphémère, une histoire dépouillée de signification métaphysique: cette absence de transcendance confirmée par la science elle-même va tender à renforcer le pessimisme d'essence philosophique présent dans l'oeuvre de Schopenhauer.... (449)

Furthermore, there is nothing in nature to indicate to Darwin that humanity as a species is eternal, or even any more likely to survive than another species; humanity is thus as much subject to the laws of nature (including the threat of extinction) as any other species of life. The contingency of human existence which will become a great motif for the Existentialists of the twentieth century may trace some of its roots back to Darwin, and certainly the pessimism of late nineteenth-century writers in philosophy and in fiction finds thematic support there.

These are the tenets of Darwin's evolutionary theory that pertain most to the French Naturalists' *Weltanschauung* in general, and to Maupassant's creative vision in particular. They are of course derived from the science of biology for the most part, and have been proposed by Darwin and his successors based on the study of evidence and extrapolation on that evidence, for the purpose of explaining an otherwise senseless morphological similarity throughout the extant world of life and in the fossil record. But Darwin's theory and influence do not end with the publication of the *Origin* or even with the work of evolutionary biologists up to this day. Evolutionary theory and what Darwin called "this view of life" have implications and effects elsewhere in the history of ideas and in popular culture. These implications and effects are also visible in the sphere of literary creation, and thus it is important to consider them under the larger rubric of Darwinism as we trace the theory's weight in the fiction of Maupassant. I refer now to that outgrowth of evolutionary theory which became known as Social Darwinism.

Social Darwinism

Subsequent to the publication of the *Origin* in 1859 and the debates it initially sparked in England between adherents and incredulous detractors, the theory of evolution took the scientific world by storm. Within a few years it had become the accepted explanatory model for the history of life; this was so in scholarly circles all across Europe and the New World (though by no means in religious circles or generally in popular culture). The notable exception to this uniformity of acceptance in academic milieus was in France, as we shall later discuss. For now, let us examine the application of Darwinism to social theory which swept across Europe and the United States in the later part of the nineteenth century.[5]

Charles Darwin was not himself the driving force behind the application of his theory to the areas of society and social policy. While he never objected to that application, he also never particularly encouraged it either. Nonetheless, he had a lot to say in his own writing about the implications of his own thought in anthropology. In his *Descent of Man* he devoted chapters to the topics "On The Development of the Intellectual and Moral Faculties during Primeval and Civilized Times" (144–165) and "On the Races of Man" 189–233). As concern natural selection operative in human societies, he wrote there,

> I have hitherto only considered the advancement of man from a semi-human condition to that of the modern savage.... With savages, the weak in body or mind are soon eliminated; and those that survive commonly exhibit a vigorous state of health. We civilized men, on the other hand, do our utmost to check the process of elimination; we build asylums for the imbecile, the maimed, and the sick; we institute poor-laws;

> and our medical men exert their utmost skill to save the life of every one to the last moment. There is reason to believe that vaccination has preserved thousands, who from a weak constitution would formerly have succumbed to small-pox. (151)

So it is evident that Darwin knew what would come of his thought where social philosophy was concerned, even if, as in this passage, questions of social policy blend in seamlessly with issues of biology. In England, the writer who picked up the standard of evolutionary thought and a form of natural selection in the context of human society was he whom Darwin called "our great philosopher, Herbert Spencer" (*Descent,* 139), even though, as Edward Caudill points out, Darwin's reaction to Spencer's thought ranged "from snide to befuddled" (Caudill, 65).

Simply put, Spencer's argument was that the principle of evolution, as effected less by natural selection than by more "Lamarckian" principles of acquired traits, is the operative rule of all nature and particularly, for his purposes, in the individual, social and moral dimensions of humanity. Spencer (1820–1903) did hold to the idea of the inheritance of acquired characteristics, unlike Darwin (discussed above), and downplayed the efficacy of simple natural selection (calling it, out of professional jealousy, an untenable hypothesis, Caudill, 71) in favor of the influence of external forces on the development of the individual. His understanding of human nature was decidedly mechanistic and non-Romantic: human nature was, for him, the collection of humanity's instincts and sentiments, which would eventually adapt to life in society. In his view, humans are motivated by a natural inclination to pursue their own self-preservation, a rational self-interest. Contrary to his mechanistic explanation of human nature, his model for society was organic, holding that social organisms were subject to the same laws of evolution as biological organisms are. The crucial point here is that these evolutionary laws provide, for Spencer, a basis for morality and human legal codes. In this vein, he argued for political and economic systems based on what he perceived to be "natural law." Hence for individuals as well as society to progress, there must be as few artificial (i.e. non-natural, not parts of nature) restrictions or interventions as possible. Understandably, Spencer was skeptical about the role of government beyond the protection of individual rights, and held for what we know as *laissez faire* policy; education, social welfare, charity, religion and care for the indigent were to be beyond the limits of government's role.[6]

During the twentieth century, in the hands of the eugenicists, Social Darwinism became a facile ideological rationale for the elimination of the weak or of those deemed undesirable or unworthy. It may seem paradoxical that William Jennings Bryan, darling of the religious conservative faction (as vividly portrayed in the *film à clé* version of the Scopes trial, *Inherit the Wind*, presenting the case for the prosecution), was in fact a Democrat and a populist: it was his abhorrence of the eugenics movement in the United States and in Germany which prompted him to champion the cause opposing the teaching of evolution; Darwinism had become linked in popular opinion with that extremist avatar of Social Darwinism, the eugenics movement.

But at its beginning, Social Darwinism had social progress as its goal, not the weeding out of the supposedly "unfit," as has been widely assumed. Social Darwinism was firmly based in Victorian times and values, including that of the belief in *progress* (differing greatly with Darwin's biology which had no such belief). Of course, another Victorian *idée reçue* was the assumed superiority of certain political orders, social classes and ethnic groups. In hindsight, we might charitably conclude that it could not help but develop into a tool for the advancement of an imperialistic, nationalistic and chauvinistic agenda. And no amount of charity could paper over the hideousness of what the eugenics movement eventually did in Nazi Germany and in the United States in the mid-twentieth century. Still, the initial aim of the Social Darwinist movement was the betterment of society, despite the fact that it was perceived to focus on the *competition* inherent in evolution's natural selection, to the exclusion of the *cooperation* which figured largely in Darwin's theory (see Caudill, 65).

Edward Caudill points out that Spencer "was a social determinist who believed that society gradually would move toward its potential in a uniform manner"(Caudill, 67). This gradualism and the notion of inexorable movement he shared with Darwin, but his belief was also one in *progress*, movement towards a goal. Spencer sought to lend his political, economic and social philosophy more credence by associating it with the science of evolution. From the mid-1860s through the 1880s, Spencer wielded influence in all areas of Western thought and among writers of all sorts, one of them being Maupassant, according to biographer Michael Lerner (165).

Important to note is the fact that Spencer's "survival of the fittest" is not a subscription to a code of competition alone, or a principle of weeding out the weak. While he did find fault with government welfare, he also supported private charity; although he (like Darwin, and Malthus before them) recognized the harshness of the natural order for the poor and downtrodden, Spencer also held for a concept of progress and made much of individuals' rights. Yet, while he did not advocate the elimination of those at the bottom of the economic ladder, he did advocate that there be some private institution, as Edward Caudill notes, "to discourage their marriage and breeding" (Caudill, 72). This attention to the dynamics of human reproduction is not lost on Maupassant and the French Naturalists. But in general, the "tooth and claw" order of competition in nature is tempered in Spencer with a concern for progress and the positive virtues of justice and generosity. Perhaps it is this admixture of the realistic with the benign that appeals to a writer of Maupassant's sense and sensibilities.

Politics in France do not seem to have delineated Social Darwinism or the application of principles of science to the field of social policy. To put it another way, there were two stripes of Social Darwinist in the latter decades of the nineteenth century. On the left, as espoused by socialists and anarchists, there was the model of Darwinism which emphasized cooperation and altruism: the struggle for life as a communal effort (Clark, 76–89). This is sometimes referred to as "reform" Darwinism. On the right, among monarchists and even many Républicains, the Darwinian model of *laissez faire* economics and Malthusian competition in environment gained favor (Clark, 90–105).

Whatever one's inclinations are in the area of social and political policy, there is one matter which must be reconciled where Social Darwinism is concerned. The simple fact is that Darwin's theory pertains to the world of biology and nature, based as it is upon observations and inference in that sphere. Social Darwinism seeks to superimpose the model of evolutionary theory on human life in society. The assumption there, and it is a large one, is that human society is in fact (and not just metaphorically) analogous to nature; the ensuing assumption is that human life in society is, or should be, subject to the same rules and principles as those which apply to life in the wild. Society must be like a jungle if Social Darwinism is the preferred model of interpretation. But already in its Spencerian origins, it makes concessions to forces which are not operative in the wild: morality, justice, and the like. Furthermore, it is necessarily selective in which aspects of biological theory to apply, even if metaphorically: Darwinian thought on variation and geographical distribution, for example, may be crucial to the overall biological theory, but have no real application where social theory is in play. In the end, Social Darwinism has always been an approximative application of general ideas, not details of theory.

Darwinism in France

Scholarship relating to the reception of Darwin in France has been made easy and enlightened by the remarkable work of three scholars, Yvette Conry, Linda L. Clark and Jean-Marc Bernardini.[7] Thanks to them the subject is readily approachable.

Yvette Conry's *ouvrage magistral* sets forth the reasons for which, on the natural sciences side, the theory of Darwin never took hold in nineteenth-century France. Whereas in England, Germany and the United States the theory of evolution had quickly been adopted in scientific circles after the publication of *The Origin of Species* in 1859, French academicians had a different reaction, and for a list of interesting reasons.

Conry first sets out a time line in four phases of scholarly response in France: 1859–1862, *mépris*; 1862–1878, *bataille*; 1878–1890, *considération*; and 1890–1900, *vulgarisation*. This is of particular interest to a study of Maupassant's fictional output, since his period of greatest productivity (from 1880 to his internment in 1891) coincide largely with the period of ebbing controversy in the world of science, when French institutions of zoology, botany, geology, paleontology, anthropology and linguistics were still mounting resistance against Darwin (Conry, 29–33). Conry also points out (and Clark and Bernardini concur) that, by the late 1880s, Social Darwinism was beginning to generate enthusiasm in France, but that all the late acceptance of the scientific side of Darwin's evolutionary theory took place in a clouded atmosphere since Social Darwinism had popularly imputed conclusions that were foreign to Darwin; thus Social Darwinism had gained a currency in popular culture during Maupassant's literary career, and that popular impression obscured the facts of scientific Darwinism (Conry, 33–34). Let us examine briefly Darwin's reception in French scientific circles.

For the most part, it seems that Darwin was long misunderstood by the French scientific community, due to prejudice, inattention, misreading, or egregiously poor translation. To begin with, Darwin was widely and erroneously assumed to be Lamarck warmed over (Conry, 40 and Clark, 19). Among botanists, the issue was misreading (or not reading) Darwin, though part of the blame may be with Darwin's choice of technical vocabulary (Conry, 165–193). Anthropology gave Darwin a particularly hard time: the influential Paul Broca misunderstood Darwin on the parallel development of analogous anatomical features, and he held the now discredited belief in multiple (in time, place and primordial forms) origins of life (Conry, 54 and 64); Armand de Quatrefages rejected Darwin's hypothesis of gradation in evolutionary forms, offering a sophist's argument that taxonomic distinctions preclude the notion of intermediary anatomical traits (Conry, 209).[8] E. von Hartmann betrayed a ludicrous reading of Darwin regarding the accumulation of anatomical changes (Conry, 59). Linda L. Clark concurs that anthropology's reception of Darwin ranged from lukewarm to hostile (Clark, 37–38). French institutional paleontology quibbled with (or misunderstood) Darwin's notion of natural equilibrium (as changeable and subject to the struggle for life), objecting that the question of extinctions clashed with their misinformed notion of natural selection's tendency to "perfect" a species (Clark, 219–220). Swiss paleontologist F.J. Pictet found fault with the utility of intermediary forms, raising the time-worn objection (answered by Darwin many times over), "what good is half a wing?"(Clark, 207) And so the story continues, with the institutions of French scholarship resisting the tide of scientific acceptance of Darwin out of a mix of conservatism, misunderstanding and scientific error.

Interesting also is the role played by other institutions of authority in late nineteenth-century France, the Christian churches. For well-known reasons (involving the metaphysical status of the human being as vehicle for the immortal soul), the Church, in the person of Pope Pius IX, reacted negatively to Darwin's theory. His successor, Leo XIII tried to reconcile faith with science, but French clergy continued to rail against evolutionary theory (Clark, 230–233). Given the Catholic Church's influence at the time, one can easily appreciate how Darwinism could be discredited by religious tradition. Protestantism and Catholicism in France even find common ground joining forces against the scientific theory (Clark, 255).

To this impressive list of forces working against the acceptance of Darwin in France, add the lukewarm reception of certain scientists like botanist A. Giard who, while claiming to accept Darwin's theory, ends up rejecting central parts of it (Clark, 240). Thus the scientific community in France, if it was not overtly hostile, was at best indifferent to the theory that was rapidly gaining currency in the rest of the Western world's intellectual establishment.

Then there is the question of the poor and even fraudulent translation of the *Origin*, the only one in French available from its initial publication in 1859 until 1872. This Clémence Royer translation misleadingly rendered, for example, the crucial term "natural selection" (with the French *élection naturelle*) leading readers to infer some mechanism of choosing implicit in nature (Henry, 264–270). Other gratuitous additions made by the translator (like her own preface, inserted for her own anti-religious or Social Darwinian agenda) and outright fabrications in the text brought about misapprehensions that were used by French anti-Darwinists against the theory.[9]

Freeman G. Henry adds other factors to Yvette Conry's long list of reasons for the French institutional rejection of Darwin. Louis Pasteur's famous disproof of spontaneous generation served as support for religious arguments against Darwinian implications on the origins of life (Henry, 294–295). Henry also asserts that French scientists saw nothing new in Darwinism, thinking that it was a rekindling of the debates between Cuvier and Lamarck in the 1830s, and not recognizing the difference between Lamarckian "progress" and Darwinian adaptation (296). There is also the very interesting point that, subsequent to the defeat of France in the Franco-Prussian War in 1870, a defeated France felt that if indeed only the fittest survive, Darwinism did not serve the national spirit well (300–301).[10] And it is a small step to the following objection, in light of the glorious tradition of French cultural *patrimoine* and the prospect of ancestry among the apes:

> How to explain divine right and absolutism in terms of that "Darwinian" heritage? What then of the French language...? How can one maintain the national myth in the face of the terrible consequences of a theory that reduces the history of mankind to biology and the history of nations to collective zoology? (Conry, 301)

Ultimately, Darwinism took the blame in France for a host of ills since it was perceived as depriving humanity of its dignity and severing it from its spiritual moorings. In some minds, Darwinism went hand in glove with a dark pessimism and a cynical view of the human condition. Certainly this is germane to the *Weltanschauung* of the French Naturalists whose fiction painted a somber portrait of life during the late nineteenth century.

Scientific acceptance of Darwin was slow in coming, and his election in 1878 as corresponding member of the botany section of the French Academy of Sciences does not by any means augur general acceptance (Clark, 10–12). The story is quite different, however, where the general public and intellectual circles outside of science are concerned. Linda L. Clark and Jean-Marc Bernardini both chronicle a widespread (if not detailed or accurate) knowledge of Darwinian principles, as well as a broad-based application of Darwinism to political, social and philosophical thought. The historiographic research they employ is impressive, and the final picture is clear: where institutional science resisted Darwinism until the end of the century and beyond, social study and popular culture welcomed Darwinian thought and models of interpretation from the 1870s onward. Gwenhaël Ponnau makes this point emphatically:

> ...à partir de la parution en 1859 de *L'Origine des espèces*, la pensée évolutionniste et ses avatars ont contribué à modeler l'imaginaire fin-de-siècle. A cela au moins deux raisons. Tout d'abord, déformée, mais aussi popularisée par les critiques virulentes et caricaturales de ses pourfendeurs, la théorie de Darwin tomba très tôt dans le domaine publique. D'autre part,...la pensée évolutionniste semble...éveiller et drainer un flux d'images spectaculaires et d'autant plus prégnantes qu'elles sont liées aux mystères fondamentaux de la vie et de la mort de l'individu et de l'espèce. (445–446)

This is the area that would seem most logical to associate with Maupassant.

Social Darwinism is the most evident manifestation of the extra-scientific community's orientation as Darwin's theory began to gain currency. Linda L. Clark argues that a permeation of American and European thought by Social Darwinism prior to 1914 has been overstated, but her rationale seems only to pertain to American understanding of the phenomenon, not the European (Ponnau, 5). The evidence that she and Jean-Marc Bernardini adduce indicates a popular and widely understood line of thought. Even if the term "Social Darwinism" was found to be uncommon in French journals and books before 1900, the term was in use in the 1880s and Darwinian vocabulary was very much in common use at that time, according to a census of the French press (Ponnau, 7). The main thesis of Bernardini's work is that, even if Social Darwinism has fallen into disrepute by 1918 in France, there was conversely a fascination (if not an *impregnation*) with it in the last decades of the nineteenth century (399). He speaks of a synthesis between positivist culture and Darwinian history that permits the foundation of a new morality, a sociology and a political science, and that this Darwinism impressed an entire generation (400). This is not to say that the "tooth-and-claw" variety of evolutionary thought was the prevailing model; we have already seen that Social Darwinism came in two varieties, one favored by the political left, and one favored by the right. The point is that a Darwinist model of thought found a reception across the political spectrum, attesting to a high degree of familiarity in society.

Darwinian thought certainly had its effect in economics. Says Bernardini,

> ...les analogies darwiniennes nourrirent pour les décennies 1880 et 1890 les themes de la concurrence économique des théoriciens de l'économie libérale; le solidarisme des années 1890 fut élaboré à partir des concepts biologiques et présenté comme une réponse à la lutte pour la vie; des lois darwiniennes furent mobilisées pour valider et défendre l'idée de concurrence et d'émulation intellectuelle d'une politique méritocratique républicaine...

and he labels a "constellation darwinienne" of social thought (401). According to Linda L. Clark, many classical economists defended the principle of *laissez faire* and Darwinian thought (62–66), and she devotes an entire chapter to the topic of Social Darwinism in sociology.

By 1890, just at the time when Maupassant's active literary career is about to conclude, there is in France what Bernardini calls a "culture darwinienne omniprésente"(Bernardini, 400). In intellectual circles outside of the natural sciences, there is intense discussion of Herbert Spencer and his variety of Social Darwinism (Clark, 38). And this applies not only to the social sciences of economics, political policy and sociology, but very much to the world of letters and journalism as well.

A young Georges Clemenceau in 1869 wrote articles in the newspaper *Le Temps* applying a Darwinist interpretation to post-Civil War reconstruction in the United States. In 1867, Charles Brooke Dupont-White published in the *Revue contemporaine* Darwinist studies of international relations (Clark, 23–24). Linda Clark reports that a considerable number of journalists and writers, as well as officials in the République, were sympathetic to Darwin and Spencer (52–62). Jean-Marc Bernardini holds that some literary authors were labeled by their contemporaries as Social Darwinians (Zola and Paul Bourget being the most often cited), and goes on to set out four categories in a census of such writers (those who use Darwinian themes, the few who borrow methodology from the natural sciences, those who develop evolutionistic philosophy to found their literary problematics or their theses, and those who set Social Darwinian characters or plots in their works) (Bernardini, 195–196).

In the institution of French education, the new evolutionary theory and Social Darwinism began to find its way in before 1880, some ten years after Maupassant has finished his formal studies. In the late 1870s, a ministerial commission is persuaded to make Spencer available to *lycée* students; in the 1880s, primary school teaching materials and books give some coverage to Darwin (Clark, 53–54).

Clark quotes Ferdinand Brunetière's somewhat sarcastic, but revealing, remark that the "habitual reader of *Le Petit Journal* or *La Petite Presse*" would appear to be "crudely informed, but nonetheless informed about pessimism, Darwinism, evolution, and the relationship of man to the monkey." (Clark, x) She also notes that eight of Darwin's books, published in French by 1880 in a series, were advertised for the general reader who would feel ashamed if found ignorant of the latest scientific controversies (29).

The case seems to be a compelling one, that, regardless of a hostile-to-indifferent reception in the natural sciences (then including anthropology), there was large exposure to a current of Darwinian and Social Darwinian ideology in the popular press, in literature, in the social sciences, in aesthetics (notably with Hippolyte Taine), in politics, a little later in schools, and generally in popular culture. It is only reasonable to assume that, with an issue of this much public interest, a figure like Maupassant would be conversant. So even if we have no direct evidence of formal education or reading in evolutionary theory, other than his biographers' reports that he read Spencer and Taine as a young man, we can safely assume that Maupassant knew well what everyone else around him knew on a matter of scientific and social controversy.

At this point, let us get a clearer notion of what it is we are about in the main part of this study. It is a Darwinian reading of the fiction of Maupassant. But what exactly does that entail? Everyone knows what it means to apply a model of interpretation (usually one derived from an identifiable ideology) to a literary corpus. We are familiar with feminist, Marxist, psychoanalytical, deconstructionist and sociocritical readings of literature; their technique involves applying ideology to the literary text to show it as a locus of thematics, characterization, action, narrative technique or structure pertinent to the concerns of the interpretive model. What is proposed for this study is an application of Darwinian evolutionary theory to the narrative fiction of Maupassant. At the outset, let it be clear that this is not an attempt to color Maupassant as "a Darwinist" or to advance the thesis that Maupassant's fiction is an organ for Darwinian (or Spencerian) ideology. Instead, it is an effort to make plain the extent to which the ideas of Darwin and the evolutionists do influence and enlighten the work of Maupassant. In light of the fact that, from the very time of its publication in 1859, the *Origin of Species* was a subject of controversy and heightened interest in intellectual and cultural circles all over the Western world, and in view of the fact that Maupassant was a reader of Spencer and manifestly interested in evolutionary theory to the extent that he was familiar with Spencer, the choice of this model for the study of Maupassant's fiction is a reasonable and promising one.

So, the "Darwinian reading" proposed would involve the application of a model of interpretation that incorporates Darwinian theory of nature and evolution (a specific understanding of natural environment, natural and sexual selection, adaptation, Malthusian "economics" of resources, heredity, success or failure to survive, and the dynamics of species) and studies from this perspective the fictional world presented in a literary corpus. Its objective is to describe the world view inherent in the creation of the corpus, but specifically in texts where the model can be profitably applied. Not all literature lends itself to Darwinian interpretation, and for that matter, not all of Maupassant does. But much of his fiction does, though, and use of the model will lay plain a logic in the texts that otherwise might remain obscure..

Why Darwin? Certain philosophical models are crucial to the world's cultural history and the history of ideas since the mid-nineteenth century. The Marxist model has long served literary study, as has Freud, though less in recent years. The point is that both have been staples in the interpretation of the French Naturalists, for the reason that they influence thought about the art of modernity to the same extent that they influenced the periods that produced it. Darwinism is a mode of thought with great implications beyond the borders of biology, and most importantly, those implications play out in the art and thought of the Western world from the 1860s on.

Let us not gloss over those implications. First, Darwin's "view of life" disarms the argument from design in nature—the assertion, usually associated with Paley, that the manifestation of "perfection" in nature, organs such as the bird's wing that are perfectly adapted to their uses, attests to the efficacy of a pre-existing designer. Darwin substitutes for design the expanse of time and the blind, slow, painstaking, accumulative process of natural selection by which the perfecting of anatomical features is buoyed to survival by the realities of the environment, supported in the bid for survival by however slight an advantage and the negative pressure on other, however slightly less fit forms.

Secondly, Darwinian theory displaces the animism of both spiritualist ontology and the alchemical model (which enjoyed currency from the Middle Ages through Romanticism), offering in its place an ideological materialism: life is no longer viewed as a metaphysical property animated by a soul breathed into the body by a creator; instead, life is now envisioned as a physical, chemical process governed by principles inherent in the physical world.

Next, evolutionary theory makes of the human mind a property of physical processes, seated organically in the body. As such, it is viewed as a product of the course of evolution. One of the effects of such an assessment is to reinforce the Darwinian notion that the species—all species—are interrelated and all derive from common ancestry by the same evolutionary processes. If human ontology before Darwin held that humankind, endowed with its faculty of reason, was partly derived from the angels, Darwin offered a new ontology that saw humankind as derived from the same ancestry as the rest of terrestrial life.

Desirably or not, Darwinism also gives rise to a metaphorical application of its principles to human society. This superimposition of methods and ideology onto social philosophy is what is commonly referred to as Social Darwinism and has taken on a life of its own since its beginnings at the prompting of Francis Galton, a cousin of Darwin himself.

Darwin's theory goes hand in hand with the new geology that made its appearance in the second half of the nineteenth century, notably with the book of Darwin's friend, Charles Lyell, entitled *Principles of geology*. In light of the new understanding of the age and variability of the earth's surface, which Darwin's evolutionary theory helps to validate and upon which the study of evolution is predicated, cosmology is necessarily and totally rewritten. Such a new development is not without its effects in the philosophy of the day. Mark Ridley phrases if concisely: "The ideas of crustal mobility and the age of the earth were, in theological terms, subversive"(10).

Finally, indeed as Daniel C. Dennett phrases it,

> But there is no such thing as philosophy-free science; there is only science whose philosophical baggage is taken on board without examination. (21)

Obviously Darwin's general effect on philosophy cannot be overstated. If, for example, Nietzsche is considered the father of existentialism, "then perhaps Darwin deserves the title of grandfather" (62). After having listed the implications of Darwin's thought in the disarming of the argument of design, the displacement of animism, the reassessment of the mind as a product of evolution, and the rest, it would be impossible not to realize the impact that Darwin's great idea would have on the thought of his and subsequent times. Darwin points forward in thought, as the implications of his theory have marked influence on human ontology. The assertion that *Homo sapiens* is merely another animal species could not have failed to elicit reaction from intellectual circles and in popular culture

Having made the case for undertaking a Darwinian reading of Maupassant, let us consider briefly the method of that reading. What will a Darwinian reading look for? What is to catch the eye of one reading a literary corpus through the lens of evolutionary theory?

First and most obviously, it would observe any direct reference to Darwin or Spencer, to evolutionary biology or to geology. Then it would give attention to the efficacy of Darwinian motifs in the text's action, thematics or narratorial vision: such motifs include the relatedness or mutability of species, natural or sexual selection, differential success in the struggle for survival or in reproduction, Malthusian environmental pressure, heredity, adaptation, and extinction.

The Darwinian model of reading would value the metaphorical representation of a fictional world as a hostile environment imposing laws of evolution on characters' lives and on history. Likewise it would seek out the depiction of nature as a Darwinist might see it; this refers to the depiction of life (especially human life, given the literary corpus) as a Darwinist might render or interpret it. Finally, this reading would take note of Darwinist logic, values or assumptions in the literary text's action or descriptions, for example, the portrayal of heredity as a determinant past.

Before effecting such a reading in the works of Maupassant in chapters three through seven, we shall consider in the following chapter "Maupassant the author," in order to establish his suitability for a reading premised on evolutionary theory.

By way of closing this chapter on the principles of evolutionary theory, I will add a word about Lamarck. As I mentioned above, this was an influential thinker in French biology in the early nineteenth century, as well as a precursor of Darwin in the matter of the mutability of species. Because of the fact that Lamarck had announced the concept of morphological change over time in species, and perhaps out of loyalty to so distinguished a figure in the history of French science, Darwin did not take the intellectual world in France quite by storm as he did in other parts of Europe and in North America. And yet, Lamarck was brushed aside when scientists came to a clearer understanding of Darwin and the differences between his proposal of natural selection and Lamarck's theory of the nature of species change.

Lamarck, not affording to the mechanics of heredity the importance that Darwin did, and not having seen through to the principle of natural selection, sought to explain species change at the level of the individual organism itself. Where Darwin proposed anatomical change by random mutation, favored in the struggle for survival by natural selection and passed on to succeeding generations by heredity, Lamarck understood change to occur in the body of the individual, in response to conditions in the environment: the most frequently cited illustration is the giraffe which, stretching its neck to reach available leaves higher up on the trees, would thus lengthen its neck by its actual feeding habits, and thus would appear the longer neck of the giraffe. The difference with Darwin is thus not a particularly subtle one, though it took a while to dawn on the French biological community, as we have seen. Darwin noted that change occurs in species when traits, already born in the individual organism and produced by random mutation, would offer an advantage to the fortunate individual in the struggle for life, and would thus aid that individual to survive longer to reproduce—and thus to pass the favored adaptation on genetically to subsequent generations. Change, for Darwin, would have nothing to do with the habits and responses of the individual in its actions, but rather with the traits with which it is born.

Darwin was familiar with the transformist theory of Lamarck, and discounted it without much thought. Perhaps if he had dealt with it more explicitly, the novelty of his own idea of natural selection and change by random genetic mutation would have been clearer to the French scientific community, hastening his acceptance immediately across the Channel. There is no direct information on the effect of the ancient and traditional mutual distrust (shall we say?) between Britain and France on Darwin's reception in the neighboring land. We can only surmise that it played a part.

NOTES

1. *The Origin of species by means of natural selection, or the preservation of favoured races in the struggle for life* (New York: Avnel, 1979) 460.
2. See Mark Ridley, ed. *The Darwin Reader* (New York, London: Norton, 1987) 14.
3. See *The Selfish Gene* (New York: Oxford UP, 1976) and *River Out of Eden: A Darwinian View of Life* (New York: Basic Books, 1995).
4. *The Descent of man and selection in relation to sex*, reprint, 2nd ed. (New York: Burt, c. 1874). References to this edition will be made parenthetically in the text.
5. Any number of good references are available on this topic. I shall rely principally on Edward Caudill, *Darwinian Myths: The Legends and Misuses of a Theory* (Knoxville: University of Tennessee Press, 1997). See also Richard Hofstadter, *Social Darwinism in American Thought* (Boston: Beacon Press, 1955) for a good general treatment of the subject.
6. For general reference on Herbert Spencer, see James G. Kennedy, *Herbert Spencer* (Boston: Twayne, 1978) and David Wiltshire, *The Social and Political Thought of Herbert Spencer* (Oxford and New York: Oxford University Press, 1978).
7. Yvette Conry, *L'Introduction du darwinisme en France au XIXe siècle* (Paris: Vrin, 1974), Linda L. Clark, *Social Darwinism in France* (University of Alabama: University of Alabama Press, 1984), and Jean-Marc Bernardini, *Le Darwinisme social en France (1859–1918): fascination et rejet d'une idéologie* (Paris: CNRS Editions, 1997).
8. One is reminded of Zeno's famous pseudo-paradox asserting that an arrow could never reach its destination since the distance it travels can theoretically be infinitely subdivided. The obvious retort, of course, is to suggest that Zeno never stood in front of an arrow. But it was Maupassant's contemporary, Bergson, who set the matter to rest in theory by pointing out that superimposing artificial delimiters (like Zeno's subdivisions of space or Quatrefages' taxonomic divides) on a continuum does not change the reality of that continuum.
9. Conry, 263–264. See Clark, 12–16 and 34–37 for a kinder view of Royer.
10. See Clark, 30 for the same point

Chapter 2:

Guy de Maupassant, "A Life"

By the time Maupassant's career of literary production ended, when he was quietly brought back to Paris in a straitjacket for internment at his own request, he had become one of the most widely read authors in Europe, second in France only to Zola. It is the consensus among his biographers that his personal demons—fascination with the supernatural, obsession with death, fear of inherited dementia, and persistent physical ailments of all sorts—led him to prefer solitude and to insist upon the strictest privacy. Toward the end of his active life he was exhibiting very erratic behavior with his family, friends and contacts in the literary world, but much earlier in life his disposition showed signs foretelling a dark turn. It remains a mystery as to how a troubled, restless mind and body could, in the concentrated span of some eleven or twelve years, produce the considerable corpus of literature that Maupassant left to the world. What is more puzzling is the breadth of thematics and mood to which that corpus attests.[1] If he was anything in life, in death and in art, he was of a character that was colorful and eclectic, given to extremes and to apparent contradictions, impossible to depict with any concision.

Born at Miromesnil (though even this is subject to some discussion[2]) on August 5, 1850, he died (of neural degeneration brought on by syphilis) on July 6, 1893 in an asylum in Passy. The legends of his life would carry rumors to the effect that his real father was Gustave Flaubert, but all of his biographers dismiss the story. His short life of almost 43 years was an extraordinarily full one. Having tasted the life of student, soldier, *canotier* and bureaucrat, he makes his way into the world of letters only in his late twenties and under the vigilant tutelage of Flaubert. He becomes one of the *groupe de Médan*, Emile Zola's circle of Naturalist authors, but before long he puts some distance between himself and the others. As Naturalists go, he clings idiosyncratically to some of the *topoi* of Romanticism, namely the *fantastique* and the exotic in his occasional use of setting (like the wilds of Corsica or the desert of North Africa, and the fiery characters who inhabit them) in a way more befitting a Mérimée than a Zola. Once his means and literary reputation are assured, he has residences in both Paris and his native Normandy, and he travels widely, ostensibly for reasons related to his health, for the most part. The pace that he sets for himself from 1880 until his final collapse is thus a frenetic one, marked by copious writing, a busy social life, travel, and a restless pursuit of peace. In his social comportment, he is unpredictable. energetic, at the same time shy and engaging, assertive yet self-conscious (mostly for the deficit in his formal education), sometimes menacing but charming as well. These are the broad lines of Maupassant's life, and their complexity does not lend much in the way of illuminating his art.

Around the time of his fourteenth birthday, Maupassant met the poet Swinburne (having been present at the poet's rescue in the waters off Etretat that summer, see Dumesnil, 74–75) who made an abiding impression on the youngster; some of Maupassant's fascination with the *fantastique* may well have had its start in his subsequent contacts with the "thin, ghost-like poet who reminded him of a sensual, perverse Edgar Allan Poe" (Lerner, 30–31). By the time he left the Lycée de Rouen and passed his *baccalauréat* in July, 1869, Maupassant had read Poe and Shakespeare (Morand, 34), but he had also read Taine, Schopenhauer, and most importantly (for our purposes) Herbert Spencer. Dumesnil reports that the young Guy was an avid reader with a rare facility for the retention of what he read (Dumesnil, 71); the reverberation of his reading will be manifest in his writing and his world view.

Poe's stylistic and thematic influence on Maupassant need not be detailed. But that of Schopenhauer is more subtle if no less pervasive. It is interesting to postulate, for example, that Schopenhauer's distinction of *phenomena* (subjective representations of the world) from reality (*noumena*)—the same distinction between the self as it appears phenomenally and the self as it is in itself—which envisions two aspects of the self, has a formative effect on Maupassant's repeated use of the motif of the *double* in his works (most notably in *Le Horla*). However, Schopenhauer's philosophy of the Will is more influential still. His view of the Will is of a blind, driving force of life without intention or finality, which manifests itself in life's struggle to survive in an unordered, godless world; thus the Will enslaves the human being and makes him or her the unconscious toy of his or her desire (Atwell, 14). Two characteristics here, pessimism and a degree of determinism, will color Maupassant's vision to a large degree.

Schopenhauer sets up an analogy of the Will/Intellect dichotomy (substituted, incidentally, for philosophy's classical Body/Soul dichotomy) by a comparison to the animal world: animal life is driven by a Will (survival and reproduction instincts, for example), and flickers of Intellect appear, culminating in the development of language in the human, but still in the service of the Will. Assuming thus a historical and organic continuum of animal-to-human life, Schopenhauer (1788–1860) prepares Western thought in general (and Maupassant in particular) for Darwinism's radical proposal of humanity's ancestry in the animal world. The French Naturalist authors, Maupassant among them, will clearly show that they are persuaded by this representation of human ontology (Magee, 94–163, 189–202, 205–225).

Bryan Magee cites Schopenhauer as the first philosopher to view the mind as a function of the physical, a physical organ at work, a survival mechanism understandable only in reference to the tasks for which it evolved. In this light, humanity does not represent a rational animal since the mind is not created for detached observation, but instead "to light the field for action"; therefore the mind is not sovereign but subordinate to the Will (263). Prior to Darwin, Schopenhauer believed that the mind evolved in nature, as a response to the necessity posed by the complexity of the world (Magee, 156).

For Schopenhauer, nature shows us that what he terms "grades" (better known as Platonic *ideals*) are individually in constant conflict: animal preys on animal, and likewise, human life is fraught with conflict. In summary, according to John E. Atwell, Schopenhauer views life as "suffering, and the source of this suffering is the will's conflict with itself."(13) Thus, a human caught up in the affirmation of Will leads a life of perpetual struggle to survive and to propagate; this current of thought flows directly into Darwinism. Furthermore, momentary happiness is transient, and at the end there is death. Lasting happiness is not in the human condition, and thus is the view of life at the basis of Schopenhauer's pessimism (Atwell, 15). This philosophy figures prominently in the reading that influences the young Maupassant for life.

Hippolyte Taine (1828–1893) was in his publishing heyday during the years of Maupassant's formative reading at the Lycée de Rouen, and Taine was a regular at the Sunday afternoon get-togethers at Flaubert's house that Maupassant attended as a young man (Lerner, 81). Taine contributed to the theoretical base of the Naturalist authors. As a group, they subscribed to his deterministic views that humankind is the product of heredity, history and the environment. His most noted dictum, announced in his *Histoire de la littérature anglaise* (1864) was that "vice and virtue are products, like vitriol and sugar." And there he outlined the three forces which, in his view, operate on an author or a school: *race* (the writer's hereditary or "racial" disposition), *milieu* (environment in the physical and the social sense), and *moment* (the writer's present time and the vector of history preceding) (Nitze and Dargan, 649–650). Again, this line of thought, with its determinism and pessimism, informed Maupassant's youth and obviously contributed to his writer's vision. Diane M. Smith identifies Taine clearly as a popularizer of Darwin's theories, who described human existence, like that of other species, as a struggle for survival (72).[3]

It would be useful at this point to take an inventory of the stock that is common between Schopenhauer and Taine, on the one hand, and Darwin, on the other. As far as theoretical bases are concerned, and to the extent that all three are "in the air" that Maupassant is breathing, all are ideologically related and very germane to this study. First there is a degree of philosophical materialism: Schopenhauer and Taine, who were at their most professionally active at precisely the time when Darwin is ruminating his overall theory (between the end of his voyage on the *Beagle* in 1836 and his receipt of Alfred Wallace's letter in 1858 prompting Darwin to rush his *Origin* to press). Schopenhauer, in philosophy, and Taine, in aesthetics and social studies, were popularizing the view of both life and epistemology as physical matters. This of course runs afoul of the spiritualism of the Church, and Darwin was to get the same treatment from institutional Christianity. In general, all three thinkers set aside the notion of God, and this certainly finds resonance in Maupassant's lifelong rejection of religion.

Second, there is the issue of evolution itself. Schopenhauer, in the realm of ideas and epistemology, anticipated Darwin and in a sense readied the ground for his emphasis on adaptation and mutation. Schopenhauer held the belief that the mind (again, the function of an organic and physical entity) had evolved in nature as a survival mechanism that developed of necessity during the history of changing life (Magee, 155–156). Darwin came upon the same idea in his observation of nature, applying it to all of life: that species (like the mind, for Schopenhauer) change over time in response to circumstances, and that some changes are favored for their helpfulness. As a general observation in this vein, Bryan Magee says that Schopenhauer

> ...prefigures many ideas that were later to be spread throughout our culture by evolutionary biology—the development of which has on the whole superseded what he had to say on these topics.... (155)

Taine, for his part, in his *De l'intelligence* (1870) described intellectual development according to a Darwinian model (Clark, 23).

Third is the determinism explicit in Schopenhauer and Taine, and implicit in Darwin. Schopenhauer's view of life is one of struggle against predetermined adversity, a struggle for survival in a hostile world, to end only in death. The thematic kinship here with Darwin should be obvious. Darwin takes the issue further by portraying a history of inexorable change, bound as history would have it for extinction. Taine's determinism takes the form of the formula, described above, of *race, milieu,* and *moment,* the circumstances which shape individual and social behavior. Similarly, Darwin's determinism also sees the past as determinant of the present and the future; its vehicle is heredity which in humanity's case is the history of our biological past and our animal parentage (Ponnau, 451–452).

In the previous chapter the point was made concerning the lowered "rank" of humanity. Philosophically in Taine and Schopenhauer, and biologically in Darwin, humanity no longer enjoys an exalted status of worth. Gwenhaël Ponnau underscores this ontological devaluation, arguing that Darwinian science stripped humankind of any transcendence and reinforced the philosophical pessimism of Schopenhauer (449). Darwinian theory clearly finds a correlative in Schopenhauer and Taine in that it makes of evolution an undirected motion, not necessarily one of progress, and which has

> ...logiquement un *terme,* une fin à l'aventure d'une humanité encore marquée par son animalité primitive et consciente de sa précarité. (446)

The animal nature of the human species is at the heart of ideological and literary pessimism that derives from Darwin. The lofty traits of humanity, like love, lose their luster and are studied for their animal (i.e. instinctual and genetic) characteristics. G. Ponnau sums up Schopenhauer's idea:

> "Le sommet des êtres"—l'homme—côtoie "l'abîme"; l'animalité est le principe même de l'amour. (451)

The literary Naturalists, Maupassant among them, with their interest in sexuality as an animal function, will seize upon this motif.

Finally, there are actual points of contact between the two philosophers and Darwin. Schopenhauer, who died in 1860 shortly after the publication of Darwin's *Origin*, became familiar with the theory of evolution very late in his life (and Bryan Magee reports that a review of the book was one of the very last things Schopenhauer read before his death). This is of course a tenuous connection, but from Darwin's side we can see that evolutionary theory built upon the materialism and determinism of Schopenhauer. Taine, on the other hand, had more to do with Darwinism than he ever acknowledged. Linda Clark tells us that Taine commented on the relationship of Darwinism to thought and society, that he makes a brief reference to the *Origin* (but did not acknowledge a great debt to it), and that Darwin supplied vocabulary to his theories whose usage is sometimes "striking." She goes on to say that Taine, in his *La Philosophie de l'art*, "made several references to natural selection, evidently considering that Darwin offered him authoritative support" (22). Clark goes on at some length to illustrate Taine's debt to Darwin, citing Taine's mission to apply Darwinism to psychology and to historiography.

The point is that Schopenhauer, Taine and Darwin are of a cloth for a young Maupassant who reads Schopenhauer extensively, reads Taine and socializes with him, and absorbs Darwin in the intellectual exchanges going on around him (as well as through his readings of Spencer). The thematic links among the three sources of ideology cannot have been lost on Maupassant, and he wove them unmistakably into his later works of fiction. Ultimately he will, in Louis Forestier's words, come to what is "...empruntée à Darwin et Spencer, la foi en un déterminisme de l'espèce."[4] The threads of Darwinism and Social Darwinism in his short stories and novels will be the matter of the following chapters.

The premise of the present study is, in a way, the synthesis of the principal findings of Yvette Conry and Jean-Marc Bernardini in their *thèses érudites*. Conry documented the French academic resistance to Darwin, describing the years corresponding to Maupassant's youth as a period of lively discussion in scientific circles of the theory of evolution by natural selection; despite this spirited debate, however, the theory was not yet a matter for educational curricula in French schools while Maupassant was a student. Full acceptance of Darwin's thought by the scientific community in France would have to wait until after the turn of the century. Bernardini, on the other hand, finds a curve of French acceptance of Social Darwinism that is almost symmetrically opposite that of Darwinian biology by Conry's account. Social Darwinism was very much in vogue among France's philosophers, sociologists and political thinkers from the 1870s through the end of the century, falling finally into disrepute around the time of the First World War's end. It stands to reason that Maupassant encountered this ideological trend in his reading during his years of *scolarité* at the Lycée de Rouen and even afterward during his brief study of law before his military service.

From this premise, it is a small step to the thesis that Maupassant's fiction was nourished by a formal knowledge of Social Darwinism and an informal knowledge of scientific Darwinism. To support this, I submit four points.

First, the biographers of Maupassant have put together a fairly complete picture of his theoretical reading during his youth. We have already seen the findings of biographers Michael Lerner and René Dumesnil who document his readings in this vein. Paul Morand as well cites Maupassant's knowledge of Spencer (83–84). Likewise Louis Forestier, in his *Introduction* to the first volume of the Pléiade edition of *Contes et Nouvelles* points out Maupassant's familiarity with Spencer and Darwin (CN1, lvii). But in all these references, it is clear that Maupassant's knowledge of Spencer comes from his readings; there is no evidence to indicate that he had read Darwin himself, and in his fiction, I infer a general—and not technical—knowledge of the principles of scientific Darwinism. In any event, it is not unreasonable to assume that Maupassant's days of theoretical reading were done in the early 1870s—in the absence of any evidence to the contrary—since he spent the rest of his life, from that point forward, furthering his literary career, caring for his family, nursing his own many ills, traveling, and writing some 300 short stories and 6 novels. Of the time Maupassant spent working in the Ministère de la Marine and onward, biographer R.H. Sherard (176 and footnote) says, "He read little or nothing" (adding that it was on account of his failing eyesight).

Second, Maupassant was a consummate socialite and *homme des salons* during his literary career; this we know in detail from the remarkable Louis Forestier *Chronologies*[5] assembled to chronicle the writer's life. We also know from the research of Yvette Conry and Jean-Marc Bernardini that Social and scientific Darwinism were both very much "in the air" as matters of interest in society, but that there were considerable misconceptions abroad regarding the scientific side of Darwinism (see previous chapter). This would seem to betoken a less than strict familiarity with Darwin's own work, leaving cultured society conversant with the generalities of the new evolutionary science, but as yet without a deeper knowledge of the subject. Such is the impression conveyed by the research of both Conry and Freeman G. Henry. In any case, Maupassant certainly knew what people were talking about in his day, and people were certainly talking about evolution. The other Naturalists were talking about it as well: Zola includes in *Germinal* a lively discussion between Etienne and Souvarine where Darwin is the explicit topic.[6] Biographer Michael Lerner tenders the following description of the social setting of the salon circle:

> ...it was all the rage to converse about Wagner, Tolstoy, D.G. Rossetti, Ibsen, and everything new and exotic from Darwinism to Angkor Wat.
>
> It was into this sparkling, hedonistic world founded on the finances of a booming industry and commerce that success and fame brought Maupassant. (192)

Third, to pursue the same tack, there is no need to recapitulate the data offered by Jean-Marc Bernardini to support his thesis, and they are considerable; they show quite clearly that evolutionary theory was a matter of great currency in the discourse of the day in late nineteenth-century France. Bernardini shows with great force of evidence that Darwinism was not far from any discussion in the arenas of science, politics, philosophy, social policy and even religion (71–285). He even includes an extraordinary appendix listing the many references to Darwinism in parliamentary debate (405–415). It is all the more unlikely that Maupassant, socialite and journalist, was not abreast of the topic informally, not having (by any biographical account) read theoretical works since his student days.

Finally, there is an abundance of signs in Maupassant's fiction itself that he was influenced by evolutionary thought and thematics. We will turn our attention to these signs in the chapters to follow.

Darwinism is an ideology, and to the extent that it finds support in the philosophy of Schopenhauer and in the aesthetics and historiography of Taine, it is all the more a system of thought. Do we draw the inference, then, that Maupassant was a theorist, a philosopher, a practitioner of a school's thought? The answer to this question, short and simple, is no. Our premise is that, while Darwinism, Social Darwinism, Schopenhauer and Taine all flow in a current of thought in the history of ideas, Maupassant was informally schooled in them, for the most part, and his extensive use of their motifs points to a proclivity rather than to a commitment to a defined creed of any kind.

Nowhere in his correspondence or in the biographies researched and written about him is there any reference to a system of thought in this (or any other) regard. Nowhere do he profess allegiance to a code of Darwinian belief, and his fiction does not really signal any systematized view of evolutionary ideas. Nor does he go to the extent of claiming adherence to a scientific (or pseudo-scientific) method in his novelistic vision of worlds real or fictional. Had he wished to, he had the perfect chance to preach the doctrine of Darwinism in particular, or scientism in general, when he appended a theoretical preface entitled "Le roman" to *Pierre et Jean*, but he did not. His grounding in these ideas remains understated throughout his writing, subtly expressed as an echo of the theory that certainly impressed him; in the words of the editor of one edition of *Pierre et Jean*, for example, in reference to Maupassant's remark that "Chacun de nous se fait une illusion du monde," Marie-Claire Ropars-Wuilleumier adds in a note that "Ce postulat subjectiviste fait sans doute écho à Schopenhauer que Maupassant admirait sans réserve…".[7] The closest Maupassant got to a manifesto was the one he agreed to write for the *groupe de Médan* (for the publicity it would gain him personally, more than for his devotion to a cause); the manifesto is more of a rejection of Romanticism's distaste for law and logic than an actual ideological or procedural creed (Morand, 82–85). Nonetheless, he does say there, "…je trouve que Schopenhauer et Herbert Spencer ont sur la vie beaucoup d'idées plus droites que l'illustre auteur des *Misérables*…" (quoted in Morand, 83).

A practitioner of a natural science, or even one who would emulate such a scientist in the observation of the world, would above all seek to be consistent, if only for the sake of preserving the appearance of the objectivity that is taken to be the hallmark of the scientific method. Even to borrow the cachet of the "hard" sciences would call upon one to maintain uniformity of premise and process for the sake of what is referred to as "control" in experimentation and observation. But Maupassant, for his part, was no scientist, and it would be misleading to imply that he was, or that he sought to be, Zola may have envisioned literary Naturalism as a something of a science practiced in the laboratory of life, but Maupassant does not seem to have plied his craft in this way. Procedural uniformity, objectivity and scientific remove are not characteristic of Maupassant's writing.

To begin with, Maupassant had a lifelong fascination with the *fantastique*. His most famous *nouvelle*, "Le Horla" and his first short story, "La main d'écorché," among others (e.g. "La Peur," CN1, 600–606 or "Apparition," CN1 780–787), revolve around thematics of the *fantastique*. This is not the stuff of science, nor is it born of an empiricist's view of the real world. In this one regard, he is closer to the Romantics whose propensities he repudiated in the *Médan* manifesto. In literature his closest influence here may have been Edgar Allan Poe, more than any Realist or contemporary. In Maupassant's literature of the fantastic, there may be an internal logic and causality, but they are not those of the Naturalist's "real" world. They explicitly defy rational interpretation, and hence explicitly disqualify the customary scientific optic of the French Naturalists. For every case like "Magnétisme" (CN1, 406–410) where the secondary narrator debunks two stories of telepathy, there are other texts in which the *fantastique* prevails. In the other short story entitled "La Peur" (CN2, 198–205), a sage character explicitly sets the *fantastique* in opposition to science, lamenting the loss of belief and the poetic to a scientistic view of the world.

A related phenomenon in Maupassant is his preoccupation with the figure of the *double*. Again, "Le Horla," among other texts of his short fiction (e.g. "Sur l'eau," CN1, 52), revolves around the mystical presence of some being who shadows the protagonist. This double exists in some textual dimension between the physical and the spiritual—certainly no sphere of existence subject to the investigation of the natural sciences. In "Le Horla" the motif of the mirror figures prominently (CN2, 935–936), as it does in "Lettre d'un fou" (CN2, 461–466)—in both cases, the mirror's reflection of the protagonist is blocked by the intervening presence of the double. In Maupassant's case, the recurrence of the motif may be a reflection of his reckoning with the haunting of mental illness, both that of his brother and his own impending difficulties. At any rate, his biographers report that he usually refused to have his photograph taken, and R.H. Sherard's book includes a facsimile of Maupassant's handwritten, firm statement of policy on this matter;[8] it may well be because of his concern with the doubling of his physical aspect. Certainly this is no rational take on the world, and the motif of the double does not draw Maupassant near to the other Naturalists in their pursuit of scientific process.

Non-scientific inconsistencies abound in Maupassant's world view. Not to make too fine a point of biography, but such inconsistencies have their correlatives in Maupassant's fiction. One example involves his misogyny. The same Maupassant who was widely known for a condescending and objectifying view of women was at the same time greatly devoted to his mother, and he showed marked respect for the women in his large circle of acquaintances who captured his admiration, such as Marie Kann, the Countess Potocka and Géneviève Strauss. These conflicting views work their way into Maupassant's writing: on the one hand, we may see a short story like "Une partie de campagne" (CN1, 244–255) with its cynical view of feminine sexuality, or "Une vente" (CN1, 1207–1212) in which a peasant farcically seeks to barter his wife, or "Le Modèle" (CN1, 1103–1109) in which a character delivers a long anti-feminine speech (1104); yet on the other hand, there are plentiful examples in the short stories and novels of strong, assertive women (e.g. "Mademoiselle Fifi," CN1, 385–397, "Histoire corse," CN1 319–323 and the same story in *Une vie*, R, 59–60), women of superior intelligence (e.g. "Bombard," CN2 365–370), or women whose plight is narrated with genuine pity (e.g. "Rosalie Prudent," CN2, 699–702 or "Clochette," CN2, 851–855).[9]

Everyone knows about Maupassant's cynicism regarding humanity. Says Albert H. Wallace,

> ...he leaves us sure that he thought nothing of the human race as a whole, but only admired its rare, unique specimens. (14)

In general, Maupassant's abiding cynicism about humanity and nature is tempered with a measure of pity. It would seem that the two would be incompatible, or that a Naturalist would not be inclined to pity the characters under the lens in his laboratory. Often, perhaps most often, Maupassant does not. But occasionally, the values of a humanitarian show through the customary roughness. The protagonist Pierre, of *Pierre et Jean*, tours the steamship where he is starting employment, and he contrasts the upper decks of the wealthy to the steerage for the poor below; his interior monologue indicates a desperate compassion for the latter (R, 828–829). Compassion is the central motif of "Misère humaine" (CN2, 752–757) and the driving narrative force in "Mon oncle Jules" (CN1, 931–938. Numerous of Maupassant's short stories show more sympathy than harshness for the characters they set before the reader. Numerous, too, are the texts which show implicit disapproval of cruelty to animals ("Coco," CN1, 1149–1152 and "L'Ane'" CN1, 902–911).[10]

Inconsistencies reveal the complexity of Maupassant's own character. He evidently enjoyed the social atmosphere of the capital, frequenting as he did the salons of Paris's high society; yet he would flee the city for the refuge of his native Normandy. His biographers report that he was shy in the company of strangers, and also that he could be imposing and assertive in conversation. He was burly and athletic, and delighted in the strenuous physical exercise of rowing, yet he made his career of the solitary, contemplative life of a writer, and was beset throughout his adult life with debilitating conditions of ill health. Serious and even somber (he often became embroiled with his professional contacts, for example), he also had a reputation as a practical joker (Lerner, 185 and Sherard, 85, 161 and 294). He showed a lifelong aversion to the practice of religion and to the Church, but one biographer reports signs that he was leaning towards an appreciation of spirituality during his last days of lucidity.[11] He was a man who, by his valet's account, cherished his pet cats, and he wrote several stories in which cruelty to animals is implicitly condemned, yet he also includes in "Sur les chats" an incident of narratorial indifference to the strangulation of a cat, and writes in *Au soleil* a confession of cruelty to a toad in his own youth (Sherard, 109–117). Biographer R.H. Sherard concludes that Maupassant "presents the most striking contradictions, and that this presents difficulties as one tries to form a valid estimate of his character (118).

Absent from his nature, then, are signs of system. True, he had daily routines, but the consistency of ideas or a coherent philosophy of life does not appear to be a part of his character. Evidently, the eclectic interior design of his residences (e.g. Lerner, 182–183) was metonymic, representative of the inner man. What we take away from all this is that Maupassant cannot be judged—or read—as an ideologue or the exponent of a system of ideas either in his life or in his works. We do not think of him primarily as a theorist, and clearly he never thought of himself as such either. Biographer Paul Morand finds that Maupassant, far from wanting the role of philosopher, actually had an inferiority complex

> ...vis-à-vis des intellectuels, devant lesquels il se sentait gêné par son peu de culture, son inhabileté à manier les idées générales et son manque d'esprit de repartie ou d'ironie méchante. (135)

Nor did he curry the favor of the scholarly community as, according to A.H. Wallace, he "stood free of the politics of academic approbation..."(Wallace, 21). There is no indication from the stories of his life that he fell into the orbit of a system of thought, or that he developed one of his own. The closest he came to doing either was with his enduring interest in the thought of Schopenhauer and Spencer whom he had read assiduously.[12]

Schools of literary practice did not interest him much more than schools of philosophical thought did. While it is true that he joined the *Soirées de Médan* and that he penned the manifesto of that circle of *les Cinq*, this professional gregariousness is uncharacteristic of him, and it appears that he joined more for the sake of advancement in the world of letters than anything else, and certainly more than out of any desire to adhere to a literary school (Morand, 82–94). In any event, it did not take him too many years to distance himself from the group.[13] This makes of him something of a reluctant Naturalist: for one whose own 1880 draft of the manifesto for *les Cinq* found fault with certain elements of the Romantic vision, surprisingly he makes occasional use of some of the *topoi* of Romanticism in his fiction; in his brief theoretical writings, he also endeavors to maintain a perspective on literature enlightened by historicism, that is, to read texts in view of their times.[14]

We get the impression that Maupassant's "Le roman" (the appended preface to *Pierre et Jean*) is much more the genuine expression of his novelistic aesthetic than his 1880 piece for *les Cinq*. By 1887, of course, his reputation and career are sufficiently established that he can speak his mind without having to adhere to the creed of a group. "Ne nous fâchons donc contre aucune théorie..." he suggests there (R, 709). With this sort of historical relativism in his aesthetics, he advises an open mind in the judgment of literary works, and situates Realism and Naturalism as simply the latest in the history of literary movements (R, 705–706). The cardinal point he makes in this preface is that Realism and Naturalism are not in themselves truth or reality, but instead the illusion of reality—and a selective one at that, aesthetically organized for artistic effect. His principle of observation (based explicitly on Flaubert), for his novelistic vision, is to observe until he can distinguish the object from all others like it, seizing upon traits that make it unique (R, 713–714). Such is the method in Maupassant's laboratory of the Naturalist, and that is as close as he gets to an empirical process of "roman expérimental," to use Zola's phrase.

In general, Maupassant rejects the notion of "schools" in this statement of his aesthetic beliefs. He defends what he calls the "poétique" as much as he does the "réaliste," for his defense of an author's right to freedom in the expression of his or her own artistic vision. Schools of criticism also get his disapproval for their exclusivity. He says,

> ...sans être des producteurs, ils sont enrégimentés dans une école, et...ils rejettent, à la façon des romanciers eux-mêmes, toutes les oeuvres conçues et exécutées en dehors de leur esthétique. (R, 704)

The moral of his preface is that literary theory should not force writers to subscribe to the rules of others. This does not sound like the precept of the exponent of any system. Biographer R.H. Sherard goes so far as to say that both Flaubert and Maupassant

> ...professed that the words "realism" and "naturalism" and other slogans of the Zola school had no signification whatever for them...

and goes on to detail Maupassant's misgivings about Zola written in a letter to Flaubert as early as 1879; the same letter also rears up at the "Naturalistic Formula" and the "human document" investigation of social conditions (157). So Maupassant, without carrying the banner for Naturalism, was something of a Naturalist by association; as Peter Whyte puts it,

> ...bien que Maupassant ait cherché de bonne heure à prendre ses distances vis-à-vis du *petit clan* naturaliste, il n'échappe pas à leur influence.... (174)

From all this we come away with the image of Maupassant as a free thinker in the aesthetics of the novel, and not the champion of a creed. There is no denying that he clearly preferred the realist's approach to writing, whether it be from the influence of Flaubert or from his proximity to the other Naturalists (or both), but it is obvious that he was not committed exclusively to the tenets of the Naturalist code, nor that he was convinced of their inherent superiority to other artistic techniques. His craft and his aesthetics were less those of a scholar than of an intelligent and talented man of the world.

From this point we arrive at the thesis for the rest of this study: Darwinism and Spencer's variety of Social Darwinism figure—sometimes prominently—in Maupassant's fiction, but it is not a scholarly application of a system of thought. I have tried to demonstrate that Maupassant was not one for doctrines of literary production, and the same attitude may be seen to hold for ideology, either biological or social. The closest he comes is his favor for Schopenhauer (himself not generally considered the founder of a *system* of thought). While Maupassant was well read in Schopenhauer and Spencer, a reader would be hard pressed to substantiate the claim that his fiction puts a system of their thought—or Darwin's—in place. My premise is that Maupassant's knowledge of Darwin, and his acceptance of Darwinism, were those of a writer educated more by common knowledge and *Zeitgeist* than by applied study. Through his reading of Spencer and, I must assume, through conversation and reading of journalism, Maupassant arrived at a sufficient knowledge of the general points of Darwinism to include it as a motif in his fiction and to use it at times as a lens through which to view the world. The forces which Darwin defined as determinant in nature may, at times, be determinant in Maupassant's work, but this should not lead us to classify this writer of fiction as "a Darwinian," lest he take on the air of an agent for an ideological system; let us also remember what Peter Whyte (174) terms the "idéologie naturaliste" concerning determinism (i.e. that determinism which directs behavior is double, one of a moral and psychological order which comes from heredity, but a second as well which is material and social, and comes from historical circumstance). So, instead of classing Maupassant as "Darwinian," let us think of him as an artist whose world view was influenced by the ideas of natural selection, the struggle for life, and the dynamics of species survival.

In the chapters that follow we will first apply a Darwinian reading to one short story, and then consider in order the general topics of Darwinian theory as they figure in the fiction of Maupassant.

NOTES

1. All references to Maupassant's works are to the Pléiade editions, *Contes et nouvelles I*, ed. Louis Forestier (Paris: Gallimard, 1974), *Contes et nouvelles II*, ed. Louis Forestier (Paris: Gallimard, 1979), and *Romans*, ed. Louis Forestier (Paris: Gallimard, 1987). For the sake of concision and clarity, references to these volumes will be noted with the following abbreviations respectively: CN1, CN2 and R, and will appear parenthetically in the text.
2. See René Dumesnil, *Guy de Maupassant* (Paris: Tallandier, 1979) 43–44. According to Dumesnil, a second *acte d'état civil* places his birth at Sotteville near Yvetot, while some say that it was at Fécamp.
3 Smith accurately identifies the influence of Auguste Comte's empiricism on Taine, as well.
4. Forestier in Maupassant CN1, lvii.
5. Forestier in Maupassant CN1, lxiii–lxxix and Forestier in Maupassant CN2, ix–xxvii.
6. Emile Zola, *Germinal* (Paris: Fasquelle, 1983) 430.
7. Maupassant, *Pierre et Jean*, Marie-Claire Ropars-Wuilleumier, ed. (Paris: Albin Michel, 1984) 22 and note 7.
8. The statement reads, "Je me suis fait une loi absolue de ne jamais laisser publier mon portrait toutes les fois que je peux l'empêcher. Les exceptions n'ont eu lieu que par surprise. Nos oeuvres appartiennent au public, mais pas nos figures. Guy de Maupassant". Sherard, 220 facing page.
9. See Sharon P. Johnson, 41–53 and 135–163 for an appropriately nuanced reading of Maupassant in this regard.
10. See Dumesnil, 212–215 concerning Maupassant's sense of compassion.
11. Sherard, 254–256. Frankly, this is a hard line to sell, especially in light of the fact that *L'Angélus*, Maupassant's last novel fragment, one of the very last things he wrote before encroaching insanity put an end to his writing career, ends in a vituperative attack on God, "Eternel meurtrier", "Eternel faiseur de cadavres", who "...a inventé le choléra, la peste, le typhus, tous les microbes qui rongent le corps" (R, 1223).
12 R.H. Sherard asserts that Schopenhauer was "one of the few authors whom Maupassant had read", 269.
13. See Maupassant, *Pierre et Jean*, Marie-Claire Ropars-Wuilleumier, ed., 17–18, note 3.
14. See, for example, "Le roman," his preface to *Pierre et Jean*, in Maupassant, R, 704–708.

Chapter 3:

A Darwinian Reading of "La Ficelle"

I have chosen for more complete analysis under the Darwinian lens one short story from among many possibilities. "La Ficelle" is a text that offers several points of interest, both thematic and narratorial, for the sort of analysis I propose. It was first published in November, 1883 in *Le Gaulois* and figured in the published collection entitled *Miss Harriet* in the following year.[1] It is a work that is typical of Maupassant for its setting in the Norman peasant society, for its narratorial perspective and for its thematics. It comes from around the midpoint in his publishing career, so it is more mature than some earlier efforts. This *nouvelle* is among Maupassant's better known short stories, often chosen as representative or simply as one of his best for inclusion in anthologies or for translation.[2]

The text opens on market day in the Norman town of Goderville as the peasants from surrounding areas are making their way in for the morning. The protagonist, Maître Hauchecorne from Bréauté, while walking into town, happens to spot a piece of string lying on the ground and bends over to pick it up. He notices that his enemy from some past dealings, Maître Malandin, has been observing him, so he hides the string in his shirt, pretends to be looking for something else on the ground, and continues on his way. Later in the day, however, Hauchecorne is summoned before the mayor and he learns that he is accused by Malandin of having found a lost wallet and not returning it to its owner. Hauchecorne protests his innocence, and the matter would seem to be resolved when the lost wallet is handed in by another peasant. But no one believes Hauchecorne's story, and he is tacitly accused in public opinion of having found the wallet and given it to an accomplice to turn in when suspicions arose. Hauchecorne becomes the object of ridicule and ostracism, and soon dies of grief, claiming his innocence to the last.

Let us begin with the level of narration. The text is narrated in the third-person voice of an omniscient extradiegetic speaker. Almost completely absent is the free indirect narrative discourse that typifies much of Maupassant's fiction, hence there is already a measure of distantiation between narrator and the characters and events narrated. It is clear from the start that this narrator sets himself apart from those characters and events, and the reader is given to understand—if only by the narrator's choice of descriptive vocabulary—that the narrator looks down on this textual society from a position of something like moral superiority. For example, his use of the term "l'aristocratie de la charrue" (CN1, 1082) to denote the clientele of the local eatery betokens an attitude of disdain and sarcasm where this society is concerned. Such downward vision lends itself easily to a judgmental tone and to the implicit moralizing that accompanies the narration of this somber action; but there is sufficient narrative distance of vision set between speaker and characters virtually to qualify as a perspective on a species entirely apart. It is as though the narrator has—and wants to have—nothing in common with the creatures whose adventures he is recounting.

It cannot precisely be said that the narrator of "La Ficelle" is taking the tone of a scientist or an empiricist in the "laboratory of life;" Maupassant does not tend to strike this pose in his writing. Nonetheless, this narrator leans a bit in that direction. Narrative distance here gives the impression of objectifying characters and events. Narrative voice becomes all the more objective, too, with the frequent use of passive voice and impersonal constructions. For example:

> Il fut confronté avec M. Malandin, qui répéta et soutint son affirmation. Ils s'injurièrent une heure durant. On fouilla, sur sa demande, maître Hauchecorne. On ne trouva rien sur lui.

And further:

> La nouvelle s'était répandue. A sa sortie de la mairie, le vieux fut entouré, interrogé avec une curiosité sérieuse ou goguenarde, mais où n'entrait aucune indignation. Et il se mit à raconter l'histoire de la ficelle. On ne le crut pas. On riait. (CN1, 1084)

The third-person passive voice and impersonal verb constructions are at times reminiscent of the language of a laboratory report. At any rate, the overall tenor of narrative discourse is one of remove in which there is no identity whatever between speaker and character.

Notice in the passage just cited that the protagonist begins to assume the role of secondary narrator. From this point onward in the text, Hauchecorne spends all his energy narrating his side of the story, hoping to clear his reputation. There is a stark difference of efficacy between him and the primary narrator: Hauchecorne persuades no one and only worsens his lot for protesting "too much;" the primary narrator, however, succeeds in communicating his objectivity (bordering on cynicism we might add) throughout, without passion if not without prejudice. Narratorial pessimism is, in this case, a link to a more scientistic view of nature and environment than we might find in the literature of another age. The implicit message is that one kind of narratorial vision is more reliable, or at least more effective, than the other.

The level of charaterization confirms what was described above. From the very start of the text, not only are all characters bereft of what we usually call "human dignity," but they are actually described as bestial and identified metonymically with their animals. The second sentence reads,

> Les mâles allaient, à pas tranquilles, tout le corps en avant à chaque mouvement de leurs longues jambes torses, déformées par les rudes travaux, par la pesée sur la charrue qui fait en même temps monter l'épaule gauche et dévier la taille, par le fauchage des blés qui fait écarter les genoux pour prendre un aplomb solide, par toutes les besognes lentes et pénibles de la campagne. (CN1, 1080)

All of the descriptive vocabulary tends toward explicitly characterizing the peasant society as beasts of burden, rendered bestial by their work. Later in the text, of course, they are rendered bestial by their actions. There is nothing surprising here: this is a *topos* of Naturalist fiction; Zola, in *La Terre* and *Germinal*, for example, made a point of similar descriptions. In "La Ficelle" the process of identification between human and animal is unmistakable.

In the second paragraph, the human masters are linked to their animals by contiguity in their movements.

> Les uns tiraient au but d'une corde une vache, un veau. Et leurs femmes, derrière l'animal, lui fouettaient les reins d'une branche encore garnie de feuilles, pour hâter sa marche. Elles portaient au bras de larges paniers d'où sortaient des têtes de poulets par-ci, des têtes de canards par-là. (CN1, 1080)

Human and beast live and work together, they go to the market together, they are so closely associated that the narrator (at least) affects difficulty in separating them or even telling them apart.

> Sur la place de Goderville, c'était une foule, une cohue d'humains et de bêtes mélangés. Les cornes des boeufs, les hauts chapeaux à longs poils des paysans riches et les coiffes des paysannes émergaient à la surface de l'assemblée. Et les voix criardes, aiguës, glapissantes formaient une clameur continue et sauvage que dominait parfois un grand éclat poussé par la robuste poitrine d'un campagnard en gaieté, ou le long meuglement d'une vache attachée au mur d'une maison. (CN1, 1080–1081)

Physical traits above the crowd and voices of people and animals are indistinguishable. The point, of course, is to mix the two to the point where humanity takes on the aspect of just another species of animal—one of the tenets of Darwinian thought. Life in the peasant society is described as revolving around association with farm animals to the point where peasants look, act, move and sound like their animals, The atmosphere of market day furnishes the pretext for the assembly of a herd of analogous species intermixed, and the narrator describes the scene as such:

> Tout cela sentait l'étable, le lait et le fumier, le foin et la sueur, dégageait cette saveur aigre, affreuse, humaine et bestiale, particulière aux gens des champs. (CN1, 1081)

"Humaine et bestiale" is the scent, bitter and awful, and it is characteristic of the rural folk. The sense of smell is added to the descriptive code that seeks obviously to portray the species as equivalent. If humans move, sound and smell like animals, the animals for their part show some degree of emotion: the fowl display a look of fear with "l'oeil effaré" (CN1, 1081). Interestingly, the narrative's next sentence begins with the feminine plural pronoun "Elles" replacing one of the two preceding feminine plural antecedents, "Les femmes" or the more proximate "leurs volailles." The reader must proceed to the verbal phrase "écoutaient les propositions" to understand whether it is human or animal "Elles" in question. Such ambiguity obviously reinforces the narrative's point that the species are, for the narrator's purposes, more similar to one another than they are distinct.

Inserted right into the middle of the presentation of the market scene is the textual introduction of the protagonist, maître Hauchecorne. Aside from the animal allusion of "corne" in his name, the effect of his characterization especially in light of what we have seen surrounding him physically and in the narrative, is obvious. He moves with great difficulty "...car il souffrait de rhumatismes" and he walks "la tête en avant, courbé en deux par ses douleurs" (CN1, 1081), so he is described as a product of, and victim of, his physical state, subject to the limitations of his body. This degree of materialism is typical of fiction written with a Darwinian slant. His physical description, his presentation interjected in the market scene, and the fact that he is observed "...cherchant dans la crotte un bout de ficelle" make him very much a part of the animalistic herd of his society. Immediately upon his introduction and his taking up the piece of string, we are told that "Il se perdit aussitôt dans la foule", blending in with that herd and subsumed by it. Nothing in his character distinguishes him from his social milieu; his only distinguishing physical attribute is his age and its attendant ailments. Again, in Darwinian fashion, this character is an individual representative of his species, not transcendent in any way, engaged in life's perennial struggle for survival in an environment that shows him no particular favor.

Far from according Hauchecorne any exalted status, the narrator seems to make it a point to present him as unattractive to the listener. "Maître Hauchecorne, économe en vrai Normand" (CN1, 1081) is first identified by his "race," that is to say by his species classification. The adjective "vrai" means something to the listener, clearly setting up a tacit bond of understanding between narrator and the narratee for whom "a real Norman" evidently should evoke a preexisting image of this social subset, and not a very flattering one at that. It is pertinent to this study that the protagonist is presented in the text first as a member of a species, set in level juxtaposition with other species, identified by class and derivation (Norman peasant), portrayed in an earthy environment and beset with physical adversity in this unfriendly setting. All that is missing at this point for this to be an overtly Darwinian conflict is a struggle for life in which the less fit perish, and such is the conflict that the text subsequently sets forth.

Self-interest appears to be the order of characterization in the text. Hauchecorne is specifically singled out as "économe" and interested in picking up any object whatever if it can serve a purpose (CN1, 1081). All the other participants in the market scene, sellers and buyers alike, are depicted as tenaciously protecting their own interests and driving hard bargains. They constitute a "foule criarde et lente, agitée par les interminables marchandages." In their efforts they

> ...s'en allaient, revenaient, perplexes, toujours dans la crainte d'être mis dedans, n'osant jamais se décider, épiant l'oeil du vendeur, cherchant sans fin à découvrir la ruse de l'homme et le défaut de la bête. (CN1,

> 1081)

The members of this species are in perpetual competition with one another, and view each other as adversaries in the struggle. The struggle is an economic one (but let us not forget that Malthus was an economist, and his notion of competition which enlightened Darwin's theory of natural selection was an economic one as well). Survival is defined in this context as scraping by financially and, on market day, not being the dupe of another's cunning in a financial transaction. Envied is he who has his own: the narrator relates, for example, what is clearly the popular opinion of the innkeeper Jourdain, "...un malin qui avait des écus" (CN1, 1082). The market place is a hostile milieu in this regard, and the peasants feel it necessary to be ever on guard lest another gain the upper hand. Resources are evidently understood to be limited, as no one is shown in an attitude other than parsimonious and vigilant. The women, for example, in trying to sell their fowl,

> ...écoutaient les propositions, maintenaient leurs prix, l'air sec, le visage impassible; ou bien tout à coup, se décidant au rabais proposé, craient au client qui s'éloignait.... (CN1, 1081)

Any human emotion that is described in these passages (apart from the "gaieté" of a peasant whose voice is made the equivalent of a cow's) is purely and solely a function of the survival instinct. Where survival is holding one's own in a milieu of limited resources, emotion is a reaction to the struggle and betokens the strain of surviving. Instinct is inbred in the species, and we clearly get that message when we are told that being "un vrai Normand" means being "économe." In this text, as it is narrated, we hear nothing about relationships of friendship (and the closest we come to altruistic or cooperative behavior is the fact that peasants are accompanied by their spouses in the market venture); we do, however, see a relationship of enemies in Hauchcorne and Malandin's rapport. We are told specifically that Malandin is Hauchecorne's "enemy" and that their bitterness stems from a financial transaction turned sour. So economics is at the heart of the action in the short story, and it provides the disposition of forces that define the social environment in which the action unfolds. Competition and adversity are the way of life in the text's circumscribed world, and the individual advances or loses by his or her ability to hold ground in the competition. Instincts play a part in the action, as for example, Hauchecorne reflexively picks up a piece of string for whatever advantage it will give him in his bid to scrape by financially. All of the forces outlined here—instinct, milieu, competition for limited resources, struggle for survival, species dynamics—are the very stuff of Darwinist theory.

We move on to the level of action where we will see that the text's central conflict works itself out on a Darwinian plan. Having set the stage very efficiently and succinctly (i.e. the first two pages establish characters, social setting and the scene of the market place), the text devotes the remaining five pages to the unfolding of action. At noon, the marketplace begins to clear out and the community retires to Chez Jourdain where they become, in the narrator's term, "mangeurs." The choice of this term is not flattering or elegant in French usage, and it has the effect of focusing on the physical process of ingesting food—this to the same extent that the meat cooking over the fire "mouillait les bouches" (CN1, 1082). Body functions, as biological rather than aesthetic in any sense, take precedence in the narrator's choice of events and descriptions to recount. Material objects can be personified in this discourse, as are the carriages and vehicles of the inn's clientele, parked in the courtyard with "...le nez par terre et le derrière en l'air" (CN1, 1082), reinforcing the impression we have of a continuum of material existence. This is one example of the materialism of the narrator's vision: from inanimate objects to animals to humans, at no point along the line is existence privileged or afforded any exalted status.

When narratorial description has ended (upon the scene of the crowded dining room, the emptying plates and jugs, the conversations about purchases, sales and crops), action begins with the crier's announcement that a wallet has been lost, followed by the police summoning Hauchecorne to the town hall. Again, Hauchecorne's physical hindrances make it difficult for him to get up and walk; it is evident that he is less than fit physically. He follows the policeman,

> ...plus courbé encore que le matin, car les premiers pas après chaque repos étaient particulièrement difficiles....(CN1, 1083)

There follows the scene of the accusation and questioning at the town hall, where Hauchecorne's word is put up against that of his enemy Malandin. The investigation gets nowhere, and legally the matter remains at this impasse for good, despite general skepticism regarding Hauchecorne's version of the story. The wallet turns up the next day, but that does nothing to exculpate Hauchecorne in popular opinion. The rest of the text's action occurs at the level of social dynamics.

From this point on in the text, a herd mentality rules social comportment and people are shown to act like animals in their treatment of Hauchecorne. Louis Forestier mentions the labyrinthine aspect of narration that comes into play at this point in the text: Hauchecorne is henceforth trapped in a cycle of endless narrating and re-narrating his story to prove his innocence;[3] in this sense, narration becomes a main motif of action. At the end of the text, Hauchecorne's cyclical narration is explicitly given the name "la Ficelle," the title of the short story. And it is with incredulity that society knowingly persecutes Hauchecorne. Society—Hauchecorne's environment—is totally without pity and indifferent to his struggle, very much in the manner of a Darwinian depiction of natural environment. Moral sense in this human milieu, if there is any at all, has nothing to do with fairness (or even giving the benefit of some doubt) to a weak, isolated figure. Strength is in numbers, and the social herd tacitly has consensus to make cruel sport of Hauchecorne's weakness and isolation.

The narrator makes an inventory of society's almost orchestrated reaction of mocking disbelief:

> On ne le crut pas. On riait.
>
> On lui disait:
> 'Vieux malin, va!'
>
> Il ne rencontra que des incrédules.
>
> On avait l'air de plaisanter en l'écoutant. On ne paraissait pas convaincu.
> Il lui semblait sentir des propos derrière son dos.
>
> Il aborda un fermier de Criquetot, qui ne le laissa pas achever et, lui jetant une tape dans le creux de son ventre, lui cria pas la figure: 'Gros malin, va!' Puis lui tourna les talons.
>
> Un maquignon de Montivilliers lui cria:
> 'Allons, allons vieille pratique, je la connais, ta ficelle!' (CN1, 1084–1085).

Note that the struggle for survival in which Hauchecorne finds himself engaged is not economic in the financial sense. Material prosperity is indeed the principal means of identification and exchange in the textual rural society, but it does not figure in the central conflict here. Money is indeed at the source of Hauchecorne's conflict in two ways: first, his feud with Malandin dates back to a financial transaction involving a horse's halter, and second, it is a wallet full of money that he is accused of finding and retaining unrightfully. But these are only the pretexts for the real conflict for social inclusion or acceptance that is, for Hauchecorne, the cause of a life-threatening crisis. So, while the society is understandable in the light of the principles of Malthusian economics, the narrative's central conflict turns out to be Darwinian in the original sense of the term. Rather than having to do with finances, it has to do instead with the economy of nature.

The struggle in question here pits the protagonist against his milieu, the individual against his species. It is not the usual Darwinian struggle of individual against peers for limited resources or for reproductive success; as was just noted, the matter leaves the resources (albeit, defined as limited in the world depicted in the narrative) aside for the remainder of the text after the wallet is returned to its owner. From that point forward, the struggle for Hauchecorne is for a commodity which, for him, is as important as physical sustenance, as it turns out. The rest of society makes it a point to deny him the credence he obviously desires, and it is obviously to fulfill a need of its own that it takes this course: as a group, they find it entertaining to induce Hauchecorne to tell his story only to greet it with ridicule and disbelief. "Les plaisants maintenant lui faisaient conter 'la Ficelle' pour s'amuser…" (CN1, 1086). No one in the social unit shows the slightest sign of awareness that this protocol is a cruel one for Hauchecorne, and it goes on unabated since society is strong enough to pursue its needs in the absence of any force—the Church, some coterie of friends, some neutral altruist, etc.—to sway the struggle in Hauchecorne's favor. As the text is written, any such force is nowhere in sight. Such a force would be motivated by a concern for justice, a feeling of pity for the weak, or simple charity—a motive or combination of motives based in morality. But the Darwinian environment of nature is utterly (even maddeningly, for us) amoral, blind to the priorities of notions like "right over wrong" or "good over evil." This is exactly the milieu that "La Ficelle" portrays, bereft of any moral support whatever for the protagonist, not even, as we shall see, from the narrator. No force, divine or natural, textual or not, will intervene to spare Hauchecorne, as he is left to survive only as best he can. The only relevant truth is not that Hauchecorne really did pick up a piece of string, but rather that only the tough are likely to survive in nature. Harsh, cruel and amoral, certainly, but it is the law of the Nature that Darwinism describes.

Tacitly, the herd detects a weakness in Hauchecorne and is willing to exploit it regardless of its effects on the survival of a lone, weak individual. Weakness is the salient characteristic of this individual: the narrative focuses on his advanced age and his rheumatism, as we have seen, but it is also made clear that the question of real innocence versus presumed guilt is one that taxes Hauchecorne greatly. His need for belief and vindication is as compelling as any other in his life as we read it. The effects of his loss in this struggle make it evident that it is a question of survival for him: he falls physically ill and soon dies as a direct result of his ostracism. The reader can only infer that one of his life needs is membership in the herd since he soon dies after being cut off from social acceptance. Meanwhile, society gets what it needs from Hauchecorne (i.e. entertainment or a target for its hostility), but the reverse is not the case.

It has already been indicated that financial economics are not the focus of our study at this juncture. The action of the text leads our attention away from money and property, and towards other concerns for the central character. The other economics, the economy of nature, is still to be addressed in a Darwinian model of interpretation for "La Ficelle."

Evolutionary biologists are often confronted with baffling phenomena which, if their theory of life is to be of use, must be understood for their evolutionary logic. Their task is to understand the economy of nature where survival strategies and struggles may not always be easily understood, and to do so they must seek to divine what advantage a species gains by being as it is. Why does a species evolve a particular trait or behavior that seems at first glance to be without sense or gratuitously unmotivated? Why, for example, would the seventeen-year locust evolve the eccentric practice of burying itself in the ground for a seventeen-year stretch, only to emerge for a short time, reproduce and dig in for another seventeen years? Why indeed?[4] There must be something gained for the advantage of the species, or else the behavior would not be "selected for" in evolutionary history. Only by crafting answers to such questions, by arriving at the internal logic of the life system, can science come to an understanding of evolution and retain evolutionary theory as a viable model for interpreting nature.

In the present application of Darwinian thought as a model of interpretation of Maupassant's short story, we are at the point where we must ask what the species (i.e. the rural society of Goderville) has to gain in ostracizing and harassing the individual who is old, defenseless and weak, but who has nothing material from which the society would stand to profit. Simple jealousy can be ruled out as a motive since there is nowhere any indication that Hauchecorne is to be envied for his wealth or lands, for example. One other plausible motive might be gratuitous nastiness, given the usual dim view that is cast upon human nature in Maupassant, but this may be set aside if we assume that species behavior is directed towards some gain that will aid in the struggle for survival; in this case, the entertainment value of leading an old man to retell his story, or bullying a defenseless Hauchecorne, does not profit the society much. So why does society seem to conspire for the suffering of this character?

If there were no answer to this question, the action would be gratuitous and the Darwinian model would come to a dead end. The absence of an answer would mean that the Darwinian reading would take us only so far and no further, after providing valuable insights up to this point into the thematics, characterization and narration of the text. The matter would remain at the point of Maupassant's customary pessimism about human nature, and the rural society of Goderville would be simply bent on persecuting Hauchecorne for the mere sake of it. Granted, that may not seem odd for a society of uncultured dolts whose transcendence and moral character are recurrently portrayed by Maupassant here and elsewhere as most lacking. On the other hand, however, the peasant society throughout Maupassant's fiction, and especially in "La Ficelle," is shown to be frugal to the point where one can reasonably assume that all conduct, all expenditure of effort, will have at its source some profit incentive—this is a plot which revolves, after all, around a character who, despite his rheumatism, bends over and picks up a string out of parsimonious obsession. So the question remains, what is to be gained by society's persecution of Hauchecorne? The answer, if there is one, must lie in the notion of the economy of nature and must shed light on the internal logic of the species in its milieu.

The solution is in the peasant society's perception and structuration of the struggle for survival in perennially hard times. Signs in the text show us graphically that success in society is measured by two standards, money and possessions, on the one hand, and standing in public opinion where finances are concerned, on the other. The tightfistedness of the characters is proof enough of the first. Proof of the second, public opinion of one's shrewdness, is evident in the prevailing opinion of the innkeeper Jourdain, for example: the narrator's choice of descriptive term reflects social opinion of this "*malin* qui avait des écus" (CN1, 1082, italics added). Society affords him respect and envy because of his business sense.

One facet of this second measure of success is the common preoccupation among the Norman peasants with not being duped (witness the description of the haggling and horse trading at market, CN1, 1081). Anyone taken for a fool will lose face in addition to losing out in a business deal, and thus will lose standing. The implication is that loss of face will bring with it a loss of respect in financial dealings, as it would be harder in the future to drive a bargain from a position of weakness. It is easy to infer that, in this peasant market economy, he who has once been the dupe has lost the upper hand in dealings.

In the action of "La Ficelle," once the wallet has been returned and Hauchecorne begins his efforts to justify himself, society makes the collective decision not to be taken in under any circumstances, objective reality of guilt or innocence aside. Society takes the position that, *if* Hauchecorne is lying, and *if* we were to let him dupe us, the shame would be on us, not on him. So, to *assure* no loss of face and no embarrassment of gullibility, the collective decision is to make Hauchecorne the victim pre-emptively, rather than risk being *his* victim. The collective thus assures itself that it will have the upper hand, even at the cost of possibly persecuting an individual who does not deserve it.

It is an economic decision, and society must weigh possible gains against possible losses. It is in the position of having to sacrifice its credence, and as a result possibly being tricked (losing) or else possibly being right if Hauchecorne is in fact telling the truth (losing nothing, but gaining nothing either). The alternative is for society to do as it actually chooses to do: withhold its belief and assume that Hauchecorne is lying (at least assuring its self-respect and dominance over the individual, regardless of truth or lies). Society weighs the certainty of its invulnerability against its putative sacrifices of truth and of Hauchecorne's peace of mind (neither of the latter weighing heavily on the collective conscience, as it turns out). Practically speaking, the only forfeit society has to make in its chosen course of action is to do without one individual who is, by society's own standards, weak, defenseless and non-productive anyway. All things taken into the balance, it is in society's interest to act as it does, given its values. That these values may be, by normal human standards, harsh and odious never enters into the question in the depiction of an amoral environment. Ultimately in the society's perception of the economy of nature, it is worth more to be assured of dominance and run no risk of loss, rather than to be vulnerable to loss or to lose. This type of economic "decision" or protocol makes sense in a Darwinian perspective, if not in an ethical one.

To validate this reading, we might ask if it is in concert with other levels of the text's composition. Where characterization is concerned there is remarkable uniformity. The character of Hauchecorne is interestingly presented as of a cloth with the rest of the Goderville society. The first sentence which speaks descriptively of him in the text, opens with the phrase

> Maître Hauchecorne, économe en vrai Normand, pensa que tout était bon à ramasser qui peut servir; (CN1, 1081).

Being a "real Norman" is to be as Hauchecorne is. Hence a "real Norman" is pragmatic and utilitarian. Society ends up putting into practice against Hauchecorne the same values he espouses. It goes so far as to include the desire to save face at all costs: in the same paragraph as the one just cited, Hauchecorne tries to dissemble as he hides the piece of string from his enemy's sight, since he is ashamed, then he pretends to be looking for something else on the ground. His salient traits here are scrupulous practicality, concern for loss of face, and deception. Obviously he is not out of place in this society, as the reader eventually finds out.

On the level of narration, the point is underscored. In one of a couple of isolated appearances of *style indirect libre* in this text, and at a particularly crucial moment, the narrator chooses to include a brutally honest thought from Hauchecorne's interior monologue, but in the narrator's discursive voice. As Hauchecorne returns home upon realizing what is behind society's treatment of him, he is

> ...honteux et indigné, étranglé par la colère, par la confusion, d'autant plus atterré qu'il était capable, avec sa finauderie de Normand, de faire ce dont on l'accusait, et même de s'en vanter comme d'un bon tour. Son innocence lui apparaissait confusément comme impossible à prouver, sa malice étant connue. (CN1, 1086)

The narrative fact is that he *could* have done it, he has the wherewithal to have done what people assume he did. The narrator knows it, and perhaps that is why he withholds sympathy or moral support for Hauchecorne as much as the peasants in the text do. At any rate, the narrator does not interject any justification for this character in the absence of the latter's ability to do it for himself. The narrator remains neutral and does not moralize; in this sense he is true to the Darwinian ground rules that prevail in the rest of the text. He strikes the pose of the impassive observer of the struggle, here an unsuccessful struggle, of an individual for survival.

Whether it is because Hauchecorne is wily and undeserving of justification, or because human nature in general does not merit the narrator's sympathy, or because that narrator is truly a detached Darwinian observer of the natural environment he has no inclination to alter, the effect is still the same. Nature is allowed to pursue its course, in spite of its detriment to the lot of the individual, and to the benefit (or at least the status quo) of the species. The entire affair is conducted in an atmosphere of amorality, and nowhere more than in the view of the narrator. As a direct result of his failure at the struggle in the economy of nature in his environment, Hauchecorne perishes: the effect of his weakness in the match against stronger natural forces is narrated as a physical effect, as he falls ill and dies as a result; it is worth noting that Peter Whyte voices his assumption of Maupassant's awareness of ideas, common among neurologists and psychiatrists, concerning the physiological origin of mental phenomena (179–180). The only extent to which the narrative leaves the domain of the material world is the extent to which the token of enchange in the economy of nature is, in this case, credence; apart from this, "La Ficelle" is a narrative of the physical world set in the environment of Norman peasant society. The forces at play are quite analogous to those which Darwin described in nature—so much so that a Darwinian reading of the text does much to bring their dynamics to light.

From this point, we proceed to a study of Darwinian motifs in the fiction of Maupassant, grouped thematically. First we will look at the premises of evolutionary theory as they figure in the novels and short stories; they are the relatedness of species, the notion inherent in Darwin's model of the tree or bush for the history of life, and the animal nature of humanity. The chapter after that will treat the core of evolutionary theory whose motifs are in evidence in the fiction of Maupassant; these motifs are natural selection (Malthusian thought and the struggle for life, survival of the fittest, and the hostile environment), the amorality of nature, cruelty and altruism, and the influence of Spencer and Social Darwinism. The following chapter will deal with sexual selection, the Darwinian dynamics of reproduction, parentage and heredity, and the final chapter will address assorted remaining topics germane to a Darwinian reading such as adaptation, random chance, and eugenics.

NOTES

1. Forestier in Maupassant, CN1, 1604.
2. For example, it appears as "The Piece of String" in Guy de Maupassant, *Selected Short Stories*, trans. Roger Colet (London and New York: Penguin, 1971) 233–241.
3. Louis Forestier in Maupassant, CN1, 1605.
4. The answer to such a question invariably comes back, "the species has hit upon a strategy that works." It is not a conscious process, nor is it directed by rational planning, but it occurs and it works. In the case of the seventeen-year locust, it is thought that the long cycle of dormancy serves to outwit predators who cannot afford to wait seventeen years for a tasty prey that they would become dependent upon and thus threaten with depletion. See Stephen Jay Gould, *Ever Since Darwin* (New York: Norton, 1977) 98–99.

Chapter 4:

Premises of a Darwinian Reading of Maupassant

Darwinian theory itself rests on the bases of conclusions drawn from detailed observations of nature. Science, for its part, has retained the theory since it has consistently been validated by discoveries subsequent to Darwin in the areas of microbiology, paleontology and geology, and also since no other theory more compelling has come along to displace it. For our purposes of studying the motifs of evolutionary theory in Maupassant (or the intertextuality of Darwinism in this corpus of fiction, as it may be put), we begin with the premises of that thought as they function on the level of motif in the novels and short stories. These have mostly, though not exclusively, to do with human ontology in the fiction; they also have to do with humanity's historical proximity to other species, as well as with the interrelatedness of all life.

The animal past in human history

We begin with the most fundamental deduction of Darwinian theory where human ontology is concerned, the notion that the historical derivation of the human species is in a parent species that was not itself human. It goes without saying that this premise of thought discounts the belief in a special, divine creation of the human species, viewing humanity instead on a continuum of life, just another twig on the historical "bush" of life's species. Just as there are remnants in human anatomy of traits of a parent species—traits like a vestigial tailbone at the base of the human spine, or the persistent recurrence of third molars (wisdom teeth) in a jawbone now too short to accommodate them—there are allusions in Maupassant's represented view of humanity to traits of pre-human ancestry.

"Le Docteur Héraclius Gloss" (CN1, 9–53) is an ambiguous *conte* from early in Maupassant's career, ambiguous in that some read it as a horror tale (e.g. Lerner, 93–94) where others may see it as an outlandish farce. In any case, it depicts the philosopher in search of the one Truth who discovers metempsychosis, Pythagoras and the transmigration of souls (a motif dear to Romanticism, to wit, the epigraph of Gérard de Nerval's "Vers dorés") During the course of his investigations, he comes into possession of a monkey which he observes to take on aspects of human comportment. Given Maupassant's general aversion to what he considered the excesses of Romanticism, we might tend to take with a grain of salt anything appearing in a text of the sort of "Héraclius Gloss." This could lead us to discount the importance of the text, were it not for the fact that a motif comes up here which is a common one in the fiction of Maupassant. Human nature is here essentially an animal nature, bereft of spiritual dimension, so the monkey which figures in the action of the text appears as a member of a species related to humanity, mimicking human behavior, donning human clothing, and so on. Consequently, the protagonist seeks to investigate "une similitude de plus avec l'espèce humaine, une probabilité plus grande de parenté," and then "...à se mettre en rapport avec la partie animale de l'humanité" (CN1, 23). The Doctor's premise is that humans and monkeys share common ancestry, and that this fact leaves the two species related with something in common in the present. It is the assumption of the latter bond (through which Gloss wants to "comprendre ce pauvre singe et...se faire entendre de lui") which contributes the oddity to this text; the assumption of common ancestry, on the other hand, underlies all of Maupassant's fiction and dovetails, as we shall shortly see, with his general view of human ontology.

Another statement of the point makes it more clearly and more credibly. In one of Maupassant's better known stories, "Les Soeurs Rondoli," the first-person narrator speaks for the text's perspective and offers a revealing description of his good friend and traveling companion. In the action of the piece, the narrator has persuaded his friend Paul to accompany him on a trip to Italy; on the train they make the acquaintance of an attractive, if unfriendly, woman who takes up for a time with the narrator to the exclusion of Paul; a year later the narrator looks up the woman's family to find that she is off with another man, so the accommodating mother fixes the narrator up with another of her daughters. The typical Maupassant representation of feminine sexuality will be of interest in a later chapter; for now, it is the description of Paul which attracts our attention. Aboard the southbound train, this friend is having second thoughts about the trip, and the narrator says of him,

> Il ressemble certainement à un écureuil. Chacun de nous d'ailleurs garde dans les traits, sous la ligne humaine, un type d'animal, comme la marque de sa race primitive. (CN2, 136)

The reader is given clearly to understand the narrator's generalized view that human nature has animal ancestry, and consequently that the human being—"chacun de nous"—has the vestiges to show for it in his or her appearance and demeanor. This betokens a belief, applying to all humanity, in a continuum of life in the history of species. In this instance where Paul is likened to a squirrel, it is less an issue of the parentage in a particular species than it is a concept of animality inherent in humanity as a function of its historical derivation.

Editor Louis Forestier points out that this is a common element in the caricatures of the time,[1] and also that the point is underscored in "Héraclius Gloss," where it is reported that the Doctor believed every animal to have

> une signification mystérieuse: il cessait de voir la bête pour ne contempler que l'homme qui se purifiait sous cette enveloppe, et il devinait les fautes passées au seul aspect de la peau expiatoire. (CN1, 22)

Apart from the slight moral implication, this heralds a properly Darwinian view of humanity descendant from pre-existing animal species.

Relationship between the human and animal species

One thematic element that Maupassant shares with his Naturalist *confrères* is the tendency to accentuate the animal side of humanity. Any reader of Zola's *La Terre*, one example from among very many, knows this thematic well. In this view of human ontology, the Naturalists make a break with the classical view of humanity as half-angel, and at the same time eschew the Romantic ontology of spirituality, favoring instead the "materialist" and Darwinian depiction of humanity as the product of biological heredity, environment and history. This brings about a portraiture of the human being as another species of animal, certainly less refined by culture than his or her counterpart in classical characterization, and that much less influenced by the spiritual and mystical than Romantic literary figures. The exception to this in Maupassant, of course, is his recurring thematic of the double and the *fantastique* in a number of his texts. This exception evokes the complexity of Maupassant addressed above, but this aside, his fiction puts very much in evidence in the majority of stories and novels the view that humanity has almost everything in common with the other species of animal life in the world. For the moment we will leave the matter of human sexuality and love for another chapter.

Characterization in *Bel-Ami* is well developed, as a function of the work's length, and the novel's central character, Georges Duroy, remains throughout a figure of little sentiment and pronounced competitiveness and ambition. His journalistic task early in the novel is to write a descriptive travelogue of north Africa, and he finds himself deficient in language to describe the long list of animal species that come to mind. He labors to give written expression to his mental images as he seeks

> ...des phrases pour bien raconter la physionomie étrange et charmante d'Alger,...l'Afrique des Arabes vagabonds et des nègres inconnus, l'Afrique inexplorée et tentante, dont on nous montre parfois, dans les jardins publics, les bêtes invraisemblables qui semblent créées pour des contes de fées, les autruches, ces poules extravagantes, les gazelles, ces chèvres divines, les girafes surprenantes et grotesques, les chameaux graves, les hippopotames monstrueux, les rhinocéros informes.... (R, 224)

Duroy's inarticulateness leaves him, for the moment, as devoid of language as those other animal species, but this is less significant than his train of thought. We notice in the same sentence Duroy's mental allusions to what are, for him, diverse human subspecies, "nous," "Arabes" and "nègres," juxtaposed with numerous animal species that are accompanied occasionally by terms of personified (and even literary) codes: "contes de fées," "extravagantes," "divines" and "graves" figure descriptively right along with "monstrueux" and "informes." All levels of life—even the "divine"—are here on the same plane where human and animal are intermixed. This point is underscored in the concluding words of that paragraph in the text:

> ...et les gorilles, ces frères effrayants de l'homme. (R, 224)

Humanity's kinship with the apes is made explicit. If the gorilla is brother to the human, then certainly the two species have common ancestry as well as common genetic traits in the present. That they share a common nature is left implicit in this passage.

In another setting, the point appears to be made a little less implicitly. The short story "Une vendetta"(CN1, 1030–1034) animal and human interact in a way as to highlight their interdependence and community of purpose. In this text, a Corsican widow is left on her own to avenge the death of her only son. As often happens in the fiction of Maupassant set in Corsica, the wild countryside stands in metonymic rapport with the savage action recounted in the narrative: here the widow Saverini patiently trains her son's hunting dog to attack a person's throat on command, playing at first on the dog's natural appetite for food. The woman goes with the dog to Sardinia, finds her son's killer in hiding, and sets the dog to its task.

Two Darwinian motifs are in evidence here: domestication and the literal notion of "tooth and claw." These are always germane to a discussion of Darwinism, and they are very much a part of the thematic outlay of this short story, though slightly removed from the point under discussion. Domestication has to do with the symbiotic relationship between two species, each serving needs of the other (though it must be said that Darwin's interest in the topic had to do with reproduction in captivity and the selection of traits for purpose). The "tooth and claw" motif, of course, relates to brutality and competition in the struggle for survival. This text, however, puts the ferocity of "tooth and claw" into the service of human needs, involving the conditioning of animal reflexes for purposes other than survival per se. Here a human protagonist interacts with a member of another species in order to effect a community of purpose, illustrating the fact that the animal has qualities useful to the human, and those qualities have nothing to do with what we generally consider humane. In fact, the human agent counts on the savage nature inherent in the beast to accomplish her ends. Given the setting, description and action of this short story, the protagonist and her dog are two facets of the same creature: the woman's ferocity leaves her only at a loss for turning instinct into action; for this, she plumbs the depths of her dog's nature and uses the animal's vitality to accomplish what she alone cannot.[2]

We note in passing that the dog here is a sort of double for the widow, and that the motif of the double (in Maupassant's literature of the *fantastique*), while not bringing other natural species into play, posits the same sort of Darwinian environment in which humanity is at odds with another species of creature, locked in a struggle for survival (or identity, or sanity, etc.). This other creature is, in other textual settings in Maupassant, literally a mirror image of the human protagonist, sharing something of the latter's nature. The hostility of the environment, even if that environment is surreal, still does call a Darwinian motif into action.

In the short story "Misti" (CN1, 1153–1158) there is an example of a domestic animal asserting itself to the detriment of a human character. The narrator, with his female companion, go to consult a fortune teller; there they notice a stuffed cat which resembles the companion's Misti. The old fortune teller recounts the story of how her own cat, many years before, had attacked her suitor who then dies of his wounds. This story leads the younger woman to give her Misti away. This is an example of hostility between the species, but motivated by rivalry for a woman's affection and attention. Competition between the species, then, can be for emotional benefit and not only for territory or environmental resources.

To pursue the topic of the relationship between species, let us consider two short stories almost identical in motif. They are "Histoire d'un chien" (CN1, 314–318) and "Mademoiselle Cocotte" (CN1, 758–763). In both texts, a coachman named François adopts a stray dog and keeps it at his employer's house until the animal becomes a nuisance, at which point the employer orders it killed; François reluctantly drowns the dog and, a month or two later, happens upon the carcass downstream. In "Histoire d'un chien" François flees the sight of the dead dog, seized by "une peur folle" (CN1, 318); in "Mademoiselle Cocotte" the action is narrated by a visitor to the asylum where François is an inmate, and the last words of the text are "Il était fou!" (CN1, 763).

The moral implication of the two texts is one that recurs in Maupassant, decrying cruelty to animals; "Histoire d'un chien" actually concludes with a tag paragraph asserting the truth of the tale and expressing hope for the success of the endeavors of the *Société protectrice des animaux*. We may reasonably infer that the narratorial position sides with this sentiment in both texts. The fate of animals is linked to the fate of their human masters, and cruelty to animals (even if not entirely voluntary or gratuitous) is not without its consequences for humans: François is driven to insane fear or to insanity in the variant versions. This points to a bond between species that goes beyond the usual understanding of domestication in which animal exists only to serve human. These short stories, on the other hand, convey the message that humans bear some degree of responsibility in their dealings with animals and that the species are interrelated as partners in life, each with the same needs, instincts and rights, if rights there be.

In another of Maupassant's better known *nouvelles*, "Yvette" (CN2, 234–307) narrative juxtaposition serves to liken human society to the society of ants described in a character's reading (CN2, 263–265). Clearly Yvette and the narrator both see the two societies constructed on the same behavioral model. What is, for our purposes, a signal paragraph in Yvette's reading is transcribed in the narrative:

> "Sans doute les singes anthropoïdes sont, de tous les animaux, ceux qui se rapprochent le plus de l'homme par leur structure anatomique; mais si nous considérons les moeurs des fourmis, leur organisation en sociétés, leurs vastes communautés, les maisons et les routes qu'elles construisent, leur habitude de domestiquer des animaux, et même parfois de faire des esclaves, nous sommes forcés d'admettre qu'elles ont droit à réclamer une place près de l'homme dans l'échelle de l'intelligence...." (CN2, 263)

In a Darwinian reading, this paragraph is fascinating. It begins with the evolutionary premise of morphological proximity of humans and apes, but goes on to strike an even closer behavioral and sociological proximity for humans with ants. This paragraph holds two mirrors to the human species, attesting to relationships of resemblance throughout the Darwinian tree or bush of life's species. Where the structure of society—insect or human—is concerned, there are issues of organization and community mentioned, as well as shelter and means of travel, but also the very issue of domestication and exploitation of partner species; all are taken for signs of analogous intelligence, but all also point to a metaphorical relationship between the species, the effect of which is to emphasize a biological, sociological and even moral equivalence among species. Of course, it is understood that we use the term "moral" guardedly, in light of the fact that we have often highlighted the amorality of the environment in the Darwinian perspective. "Moral" here is used in the anthropocentric sense where "human" characters interact with the world around them, and a "human" narrator makes implicit judgments of their behavior. In any event, the point is that the metaphorical relationship described here among species attests to resemblance at all levels: anatomy, behavior, needs, instinct, social structure, and so on.

The rapport between species can take the form of human empathy, as it does twice notably in *Mont-Oriol*. First, there is the scene in which there is to be the blasting of a large boulder at the future resort, and after the fuse is lit a stray dog is spotted near the demolition site by the crowd of spectators; the central character Christiane is moved by fear and pity for the dog and her would-be lover Paul rushes down to the site to attempt a rescue. The dog evades Paul who is forced to abandon the animal to its fate (R, 505). Christiane, "indifférente maintenant" where the dog is concerned, draws favorable conslusions about Paul's passionate nature (R, 506), but while leaving the area she realizes that she has trod upon some of the bloody remains of the dog, at which point

> Elle suffoqua, tellement émue qu'elle ne put retenir ses larmes. Et elle murmurait en s'essuyant les yeux avec son mouchoir: "Pauvre petite bête, pauvre petite bête!" Elle ne voulait plus rien entendre, elle voulait rentrer, s'enfermer. Ce jour, si bien commencé, finissait mal pour elle. Etait-ce un présage? (R, 508)

The bloody demise of the dog is, for Christiane, a bad omen, as she senses her destiny to be somehow linked to that of the dog.

Then, later in the novel, Christiane feels sympathy for a donkey which has been worked to death and left in the road where it fell. She is "navrée, bouleversée" as she imagines the life of the animal, first a "petit bourricot joyeux" which was then pressed into unremitting servitude:

> ...l'incessante et terrible marche par les interminables routes! les coups! les coups! les charges trop lourdes, les soleils accablants.... (R, 647)

And Christiane's thoughts of pity go on for another half page, at which point the narrator, through Christiane's point of view, describes the human family which, as it turns out, had abandoned the donkey's carcass; it is a desperately poor family, "une famille humaine!" about whom Christiane concludes, "Ils mourraient probablement comme était mort leur bourricot." (R, 648) It seems that, in this world of the novel, humanity cannot extricate itself from the common destiny of all animal species.

The relationship between species can be detrimental to one or to both in the worst of cases, but it can be a beneficial symbiosis, too. This aspect of cooperation can even be accompanied with a measure of humor in Maupassant. "Toine" (CN2, 426–4435) relates the story of a gregarious innkeeper who, when paralyzed by a stroke, is pressed by his wife to incubate eggs with his body heat—an arrangement which ends up working out well for all parties. Species can profitably depend on each other for help in the struggle for survival, and this applies to humans as much as to other species: in Toine's case, he and his wife profit from the increase in their chicken population, Toine finds a way to be productive in his paralysis, and he is able to appease his irascible wife; the chickens, for their part, would not survive without the artificial incubation. We will later have occasion to look at the general notion of cooperation in Maupassant's Darwinian thematics.

Descriptive juxtaposition of humans and animals (assimilation)

There are numerous cases in Maupassant's fiction of the effective juxtaposition of human characters with animals in descriptive passages; the effect in question is the assimilation of the two in the narrator's world view. A most graphic example of this phenomenon has already been detailed in the preceding study of "La Ficelle," in the opening description of the morning market where peasant and livestock intermingle and are indistinguishable in sound and appearance.

A very similar setting figures in "Les Sabots" where, at the outset, the narrator makes the assimilation explicit. The text opens on a church service whose peasant faithful emanate on a July morning "une odeur de bétail, un fumet de troupeau" (CN1, 711) Voices of roosters and cows enter through the open door, as do the smells of the neighboring fields. Then the narrator makes a clever pun, referring to the priest's remarks to his *ouailles* (ibid.). The peasant parishioners are "sheep" in more than a figurative sense. Here and elsewhere in Maupassant, the society's lifelong association with livestock brings about a metonymic relationship between humans and animal species in the narratorial perspective.

"Le Baptême" is another short story set in the same Norman peasant society context, and it proceeds from the same assumptions on human ontology as all of Maupassant's writings cast in this social setting. The text's second paragraph is a fairly unattractive description of a sow dozing

> ...sur le bord du fumier, le ventre énorme, les mamelles gonflées, tandis qu'une troupe de petits porcs tournaient autour.... (CN1, 1144)

Then, shortly after a brief description of the farm setting, its ducks and some of its smells and human inhabitants, there is introduced

> ...une grosse femme...qui portait un enfant de deux mois...[qui] reposait sur le ventre en bosse de la garde.... (CN1, 1145)

The proximity of the two descriptions can leave little doubt as to the implications that this narrator signals.

The text entitled "Le Fermier"(CN2, 814–821) sets the same tone, and its descriptive language puts the matter in a historical dimension. The added implication here is that ontology is a function of the past. The Norman peasant cottage is situated in "une de ces cours à pommiers" with which the narrator assumes his listener is familiar. Descriptive vision focuses on the house, again in some measure metonymically for its inhabitants:

> On sentait, là-dedans, le lait, la pomme, la fumée, et cette odeur innommable des vieilles maisons paysannes, odeur du sol, des murs, des meubles, odeurs des vieilles soupes répandues, des vieux lavages et des vieux habitants, odeur des bêtes et des gens mêlés, des choses et des êtres, odeur du temps, du temps passé. (CN2, 815)

At the beginning of this sentence is an approximation of the ancient formula animal / vegetable / mineral: *lait* (animal product), *pomme,* and *fumée* (followed by *sol, murs* and *meubles*), the effect of which is to reduce all material existence to a common denominator. Then there are four mentions of the adjective "old," concluding with the phrase *temps passé.* The materialism and historicism at play here are well suited to the overall Darwinian world view. But the phrase that crystallizes the pertinent view of species is "odeur des bêtes et des gens mêlés, des choses et des êtres". The assertion is one with which we are by now familiar. Humanity, as another animal species, is bound up in existence with its analogous biological relatives in a material world that has been formed by its history.

Belgian peasants are treated to the same descriptive lot in "De Paris à Heyst," a chronicle of Maupassant's hot air balloon ride northward from Paris. Upon landing in Belgium, the traveling party is greeted by an incredulous ensemble of men and beasts, described in the customary fashion:

> Puis, en même temps que les hommes effarés, dont quelques-uns sautaient d'étonnement avec des gestes sauvages, toutes les vaches qui passaient sur les dunes venaient à nous, entourant notre ballon d'un cercle étrange et comique de cornes, de gros yeux et de naseaux soufflants. (CN2, 1290)

Human and animal display the same reaction to the surprise visit, wordless, wide-eyed and even "sauvage."

It should not be inferred that Maupassant reserves this view of human ontology for the peasant class. For the relatively privileged Jeanne, the central character of *Une vie*, the technique of descriptive assimilation of species is more subtle, but no less effective, than that in the Norman peasant short stories. Early in the novel, the description of the salon in Jeanne's family home suggests a metonymic association of humans and animal figures in the décor, and then uses an intertextual reference to the fables of La Fontaine to underscore the point:

> Au rez-de-chaussée, à droite, on entrait dans le salon démesuré, tendu de tapisseries à feuillages où se promenaient des oiseaux. Tout le meuble, en tapisserie au petit point, n'était que l'illustration des *Fables* de la Fontaine; et Jeanne eut un tressaillement de plaisir en retrouvant une chaise qu'elle avait aimée, étant tout enfant, et qui représentait l'histoire du Renard et de la Cigogne. (R,9)

The allusion to La Fontaine is particularly apt, but not for the classically understood reason of personifying animals to illustrate the loftier human condition; rather it has the inverse effect in Maupassant of reducing humanity to the level of biological species. In other words, animals can be portrayed as having the same issues to deal with as humans, and then, vice versa. The net effect is that the species are on a par with one another.

Physical traits of the animal in humans

The physical attribute of the animal in human characters is fairly commonplace in Naturalist fiction. Speaking of European Naturalists in general, Diane M. Smith holds that

> Characteristic of these novels is the use of animal imagery and other metaphors of evolutionary regression to depict the dehumanizing effects of poverty. (76)

Maupassant takes the motif beyond the thematic of poverty alone. Rarely is it a flattering motif for the human character; instead, it is a means of reminding one of the animal side of the human condition. We have already seen one instance of this above, in the discussion of "Les Soeurs Rondoli" where the narrator says of his friend Paul, "Il ressemble certainement à un écureuil" (CN2, 136).

We return to *Une vie* to see the case of the mayor's wife. The reader is not surprised to find such a description of a character accompanying a scene of near farce, the blessing of a boat and the attendant procession to and from the event. Jeanne takes the arm of the vicomte Julien, soon to be her fiancé and negligent husband, and follows priest, altar boys, *serpent* player and three singers who "détonnaient à pleine gueule dans la claire matinée" (we will pass on the dead metaphor of *gueule* as an allusion of animal characteristic, only to note the humor of the procession's ersatz solemnity) (R, 33). During the proceedings, Jeanne signals her acceptance of the vicomte's marriage proposal, and the procession makes its way back into town from the shore.

> Le retour fut une débandade. La croix, entre les mains de l'enfant de choeur, avait perdu sa dignité; elle filait vite, oscillant de droite et de gauche, ou bien penchée en avant, prête à tomber sur le nez. Le curé, qui ne priait plus, galopait derrière; les chantres et le serpent avaient disparu... Une même pensée, qui mettait en leur tête comme une odeur de cuisine, allongeait les jambes, mouillait les bouches de salive, descendait jusqu'au fond des ventres où elle faisait chanter les boyaux. (R, 34)

The priest "gallops" and everyone is bolting for the restaurant where lunch is waiting; the terrestrial side of life is, for the moment, trumping the religious stateliness of the occasion and contrasting with the expected loftiness of a religious ceremony. The music of the *serpent* is replaced by the growling of people's stomachs. The animal side of humanity is clearly taking over.

At this point in the narrative, the scene shifts to the restaurant where the party is at table. Jeanne's father is flanked by the mayor and the mayor's wife upon whom the narrator concentrates for a moment.

> Elle avait une figure étroite serrée dans son grand bonnet normand, une vraie tête de poule à huppe blanche, avec un oeil tout rond et toujours étonné; et elle mangeait par petits coups rapides comme si elle eût picoté son assiette avec son nez. (R, 34)

This woman not only looks like a chicken, but acts like one in the way she looks around and eats. And her resemblance to a hen is underscored by the emphatic "une vraie" (just as Paul's resemblance to a squirrel, in "Les Soeurs Rondoli," is accentuated by the emphatic "certainement"), as if the narrator wishes to insist on the character's likeness to the animal in question. The narrative code of description through the entire scene, then, is colored by a vocabulary and motif of the animal. In the case of the mayor's wife, it is not only in physical attribute but in behavior as well.

The well-known *nouvelle* "La Maison Tellier" (CN1, 256–283), the story of the temporarily closed brothel, deals as a matter of main motif with sexuality as rather an animal function, and in the process makes the same chicken comparison. The women of the Maison Tellier, on a train from Rouen to Virville to attend the first communion of the Madam's niece, are joined in the compartment by a woman in her "toilette rustique" carrying a large basket that "...laissait passer les têtes effarées de trois canards" (CN1, 263–264). The ducks share some degree of identity with the woman who is described immediately thereafter as having "...une physionomie de poule avec un nez pointu comme un bec" (CN1, 264). The ducks become the objects of humor and attention for all the other passengers in the compartment, and permit the opening of conversation between the Tellier women and a traveling salesman "aux favoris blonds"; the peasant woman and her husband themselves never speak, but shortly before they disembark, they—like their poultry—become in their turn the object of humor, and it is because of their resemblance to chickens:

> Les deux paysans. figés dans l'ahurissement, regardaient de côté, d'un seul oeil; et ils ressemblaient si absolument à des poulets que l'homme aux favoris blonds, en se relevant, leur fit dans le nez "Co-co-ri-co". Ce qui déchaîna de nouveau un ouragan de gaieté. (CN1, 266)

Again, the narrator underscores the human-animal resemblance with an emphatic, here *si absolument*. The effect of the similarity of species, for the human characters, is humiliation. So if there is anything to be extrapolated from this episode (and the others like it), it is that the depiction of humanity in Maupassant is, for humanity's proximity in appearance and behavior to other animal species, not calculated to invest the human species with dignity, but instead to point up its consubstantiation with the animals. And it does so invariably in a derisive narrative tone.

In another text, already comical for its subject matter of a preposterous case before a Norman court of civil law, we find described a setting by now familiar to us. The narrative of "Tribunaux rustiques" (CN2, 388–392) opens by presenting the "paysans" waiting for their day in court:

> Il y en a des grands et des petits, des gros rouges et des maigres qui ont l'air taillés dans une souche de pommiers. Ils ont posé par terre leurs paniers et ils restent tranquilles, silencieux, préoccupés par leur affaire. Ils ont apporté avec eux des odeurs d'étable et de sueur, de lait aigre et de fumier. (CN2, 388)

As a customary part of their personae, the members of this society have about them the odor of the animals next to whom they ostensibly spend the greater part of their lives. As a function of this lifelong juxtaposition, they take on a physical characteristic of the other species. The narrator clearly seeks to emphasize this assimilation of trait, as it is reported:

> Et sur la muraille nue, un grand Christ de bois, tordu dans une pose douloureuse, semble offrir encore sa souffrance éternelle pour la cause de ces brutes aux senteurs de bêtes. (CN2, 388)

The people to whom this narrator refers are "brutes" by designation, and they have the smell of their barnyard animals permanently about them as if it were their own. In the narrative description, the peasants are as if carved out of apple tree trunks (i.e. vegetal trait), and they resemble their animals in odor (animal trait); these two characteristics are offered in the place of anything properly "humanizing," making the point certainly by omission that these characters have little separating them in physical attribute from the other species that inhabit the Norman countryside.

Character traits of the animal in humans

If the assimilation of physical traits of animals is common among Maupassant's human characters, the assimilation of animal traits in behavior or personality is even more common. To be sure, the human characters articulate and reason and love, and the novels especially develop characters' personalities and detail their faculties. But, and this is the crucial point, the animal side of their nature is never far off. Even in the human faculty of love, a very positive character like Elisabeth in *Notre coeur*, with whom the reader can easily empathize, tenders her love for Mariolle "animalement" (R, 1170).

The principle applies to characters from across the social spectrum. Of course, the peasants depicted in the novels and especially the short stories are thus described most often, but city dwellers and socialites and members of the higher social classes are in for their fair share as well. Josephine, the Roland family's maid in *Pierre et Jean* "...possédait à l'excès l'air étonné et bestial des paysans," (R, 727) but the mayor Renardet (as his cratyllic name might suggest) is motivated by "un emportement bestial" (CN2, 639) as he assaults the title character in "La Petite Roque." Throughout that text, the mayor, who otherwise is a pillar of his community, is imbued with animal characteristics: he has "l'instinct mystérieux et confus qui guide tous les êtres en danger" (CN2, 639); ",,,il dormit d'un épais sommeil de brute..." (CN2, 640); of a "nature brutale [qui] ne se prêtait à aucune nuance de sentiment ou de crainte morale", he is "plein d'instincts sauvages"(CN2, 640) and

> ...il avait gardé au coeur, gardé dans sa chair, gardé sur ses lèvres, gardé jusque dans ses doigts d'assassin une sorte d'amour bestial.... (CN2, 641)

Then there is the wealthy Paul Brétigny in *Mont-Oriol* of whom the narrator says, "Il aimait la campagne avec ses instincts ardents où transperçait toujours de l'animalité." (R, 533)

In "L'Homme de Mars," (CN2, 1003–1010) the narrator's eccentric interlocutor confesses that "...je suis une bête à mon tour" and goes on to aver that "...notre intelligence...dépasse à peine l'instinct des animaux" (CN2, 1005). This opinion is offered by way of a general assessment of humanity, again defined as being on a par with the animals, but even where the vaunted faculty of reason is in question.

Another generalization about human nature appears early in "Le masque" (CN2, 1134–1142), a pessimistic narrative about human nature and aging. In this short story, an older man cannot come to grips with the passing of youth, so as often as possible he dons a mask and goes to costume parties where he dances wildly until he passes out from exhaustion. In the first paragraph, the narrator posits as a general fact "...cet irrésistible désir de sauter, d'avoir chaud, de s'amuser qui someille au fond de l'animal humain" (CN2, 1134). It is the "party animal" which he speaks of in all of us, but the human being nonetheless is called the "human animal." As if to reinforce this dehumanizing description, the narrator describes the party crowd as "La haie onduleuse, la pâte remuante de femmes et d'hommes..." (CN2, 1134). For this narrator, then, the human desire to socialize in an atmosphere of music and revelry is an animalistic one.

In a novel action thematically related to the motif of "Le masque," Bertin of *Fort comme la mort* is a protagonist whose general malaise over the loss of his youth exacerbates his jealousy in love. This emotion of jealousy, a purely human one, is described in the character's narrated point of view as "...une envie animale de hurler à la façon des chiens attachés," (R, 987). It is obviously the case that, in Maupassant's overall view, even the emotions are an animal function, and this is very much in keeping with a generally Darwinian, materialistic view of life and the world. Love does not give a character to transcend in Maupassant as it had done in the literature of the Romantics. Rather it is another trait of identity with the other animal species, if indeed its power and impetus constitute an "animal desire" as it is depicted here and often elsewhere in the fiction of Maupassant.

The characteristic of the animal is directly ascribed to humans often in Maupassant's narrative discourse, most frequently without any other qualification or further explanation, as if to imply that such designation stands on its own without need for elaboration. The servant girl Rosalie in *Une vie,* very shortly after her first introduction in the novel, is shown daydreaming in the family carriage during a trip:

> Rosalie, un paquet sur les genoux, songeait de cette songerie animale des gens du peuple. (R, 6)

We note in passing the famous Maupassant demonstrative, *cette,* which makes an implicit appeal to knowledge shared (and generalizations assumed) by the narratee. Here the narrator presumes that his listener is familiar with a characteristic of "the folk," and in this case the characteristic is an animalistic daydreaming; the description has the effect of making something of an animal of Rosalie since she displays an animal's behavior to some degree.

In fact, throughout the novels and short stories, human needs, tastes, functions and emotions are invested with the descriptive code of the bestial. The descriptive term in question is usually mentioned in isolation and not further developed thematically, as if to say that the point is accepted and assumed by narrator and listener in a tacit agreement of understanding that postulates the bestial nature of the human subject. The "Vagabond" Jacques Randel suffers "une faim de bête, une de ces faims qui jettent les loups sur les hommes" (CN2, 857); in the subsequent action of the text, Randel does hurl himself somewhat at human society in his desperation. In the process, as he squares off against the mayor of a village along his way, the two of them "...se dévisageaient tous les deux avec une haine de bêtes appartenant à des races ennemies" (CN2, 863). In "Le Lapin" the peasant wife is described succinctly as

> ...une grosse fille très rouge, joufflue, dépoitraillée, ventrue, large des hanches, une sorte de femelle sanguine et bestiale, la femme du berger Severin. (CN2, 971)

The reveling patrons of the brothel in "Le Port" are shown in their shouting and drinking, as each one "...lâchait en liberté la brute humaine" (CN2, 1128). Descriptive simile plays the same role of qualifying the human with animal characteristic in "Le Champ d'oliviers" where the priest's long lost son turns up and evokes the narrator's description in *style indirect libre* as one "qui sentait le vice comme un bouc sent la bête" (CN2, 1189).

For Maupassant narrators, revelers readily bring out the descriptive code of bestiality. The floating restaurant, *la Grenouillère* (evidently a real-life establishment along the Seine west of Paris where Maupassant himself spent a considerable amount of time during his *canotier* days), figures in two short stories, and in both cases the clientele garners descriptions in bestial terms. In "La Femme de Paul" we read that

> Les hommes...s'agitaient en vociférant par un besoin de tapage naturel aux brutes. Les femmes, cherchant une proie pour le soir, se faisaient payer à boire.... (CN1, 294)

Then, in "Yvette," again at *la Grenouillère*, we see similarly that

> Les buveurs...vociféraient sans raison, cédant à un besoin violent de faire du tapage, à un besoin de brutes d'avoir les oreilles et le cerveau pleins de vacarme. (CN2, 265)

The *nouvelle* "Allouma" (CN2, 1095–1117), a framed narrative in which most of the text is spoken in the voice of Auballe, a French colonist in Africa, bears repeated witness to the belief in the animality of the human being in this (decidedly chauvinistic) character/secondary narrator's view. In this work, Auballe recounts to the primary narrator, a visiting friend of a friend, the story of his obsession with an Arab woman who has finally left him for good. Auballe's biases are patently clear in his descriptions throughout his narrative. He qualifies the face of Allouma, the object of his infatuation, as "un peu bestiale" even though only half a page before he had referred to her as "une fille au visage d'idole" (CN2, 1101). Hers is "le sourire animal de ses lèvres" (CN2, 1102). But he does not exclude himself from the descriptive code of the animal: he speaks of their first encounter as

> ...une lutte courte, sans paroles, violente, entre les prunelles seules, l'éternelle lutte entre les deux brutes humaines, le mâle et la femelle, où le mâle est toujours vaincu. (CN2, 1102)

Allouma is "souple et saine comme une bête, avec des airs, des mouvements, des graces et une sorte d'odeur de gazelle" (CN2, 1102). For Auballe, Allouma is "...cette créature d'une autre race, qui me semblait presque d'une autre espèce," and he declares that he hadn't actually loved her, since he finds that Arab women are "...trop près de l'animalité humaine" (CN2, 1107); he evidently changes his mind to some degree, as he says in the end,

> Je l'aimais en effet, un peu comme on aime un animal très rare, chien ou cheval, impossible à remplacer. C'était une bête admirable, une bête sensuelle, une bête à plaisir qui avait un corps de femme. (CN2, 1114)

But again, he says of himself, "...je fus attendri comme une bête" (CN2, 1112), though it does not go very far toward absolving him of some odious prejudices which are, nonetheless, grounded in a belief in the proximity of animal and human species.

Animals with human traits

There are cases of the relationship opposite to that described above, that is to say, cases of animals depicted in Maupassant as having human traits. In the short story entitled "Amour" (CN2, 845–850), the title refers not too metaphorically to the relationship between two mates of a bird species. In the action, the narrator kills the female of a bird couple while hunting with his cousin. The male then cries out for his mate and will not leave the site despite the danger to him; the cousin ends up killing the male bird. Clearly, and especially in light of the text's title, the point is to portray emotion normally reserved for human character as being at home in other species. It is interesting to note in passing that in this text there are two other Darwinian references made manifest: first, the narrator states his love for swamps since the environment there represents an entire world of life; secondly, there is expressed a very Darwinian hypothesis about the beginnings of life in what biologists call the "primal soup" of a swamp (CN2, 846).

Another example of the assignment of human characteristic to an animal is the curious case of the parrot in "Le Noyé" (CN2, 1038–1044) where a fisherman who is abusive to his wife drowns in a storm, and his body is never recovered; the woman buys a parrot and begins to hear her husband's abusive voice in the house; she thinks it is the husband returning, and when she discovers that it is the voice of the parrot, she kills the bird. Granted, this is closer to Maupassant's *fantastique* strain, but the bird does manage to take on a physical trait of a particular human character. Like the dog in "L'Auberge" (CN2, 784–796), inadvertently left outside to freeze, the animal has a voice that comes to haunt the human master in eerie fashion. We have also already mentioned the dog in "Une vendetta" (CN1, 1030–1034) and the cow in "Le Vagabond" (CN2, 859–860) which serve so well the needs of human characters that the latter show a bond with the animals.

Metonymy with wild natural setting

At this point, let us bring up the matter of setting and its metonymic rapport with action in certain of Maupassant's fictional texts. Short stories and episodes of a novel which are set in Corsica exploit the savage nature of the countryside to the effect of reflecting the savagery of the action and of the human characters who perpetrate it. For the purposes of this study, the effect is also to make of humanity the product of its environment, responsive to the conditions surrounding it.

There is an episode early on in *Une vie*, during the honeymoon trip in Corsica, where the couple meets a young woman who, when offered a gift by the couple, requests that she be sent a pistol with which to kill her brother-in-law (R, 60); the same plot motif figures in the short story "Histoire corse" (CN1, 323). In these textual passages, the wild surrounding terrain mirrors the savagery of the characters who inhabit it; nothing could be more obvious in "Un bandit corse" (CN 1, 436–439) as well.

It is not an exaggeration to say the same about the jungle atmosphere of the city depicted in some other places in Maupassant, as in *Bel-Ami*, for example. The harshness of the economic and social setting brings out the animal in a character like Georges Duroy. In this case or in the others just described, the phenomenon of metonymy is purely literary, but the function is to make of the environment a force to which human subjects respond by imitation: they incorporate the rules of the environment in their conduct; thus the environment is a model for action, and characters reflect it in their own nature.

Relationship between individual and species

In Darwinian theory, the relationship between an organism and its species is a matter of close study. The whole notion of evolution revolves around resemblance to (and membership in) a species, and departure from the norms of its morphology and behavior in the process of delineating new species. The question of rapport with the species becomes a matter of thematic interest occasionally in Maupassant. In "La Tombe" (CN2, 213–217), the tale of a lawyer on trial for digging up the grave of his beloved out a need to see her again, the accused offers the following justification in his own defense:

> Et pourtant il en naîtra des milliers de créatures, des millions, des milliards, et bien plus encore, et parmi toutes les femmes futures, jamais cella-là ne se retrouvera. (CN2, 216)

The genetic uniqueness of the individual is stated as fact, even in the context of biological generation and the astronomical numbers that it implies.

As we saw in the case of "La Ficelle" studied in the previous chapter, the individual can become the victim of the species, for reasons important to the species alone. A similar development comes about in "Une famille" (CN2, 763–768), ironically titled to point up the occasional hostility with which the "species" can treat the individual. A family is generally understood to be a protective and nurturing social unit, but in this short story it is a circumscribed arena of persecution for a debilitated grandfather whom everyone teases cruelly with the foods he likes. Here the species in question serves its need for entertainment in what the visiting narrator describes as an uninteresting provincial town.

"Le Vagabond" (CN2 856–867) also depicts the individual as put upon by his species, here as the outcast who serves the species' evident need for a scapegoat to look down upon. In this instance the treatment has the undesirable effect of inducing the subject to live down to society's expectations. The individual is now forced to live on the margins of society by society's choice to exclude him from its economy. As tends to happen in Darwinian geography, life on the periphery of the species' range can adapt by change of survival strategy, and in "Le Vagabond," Jacques Randel evolves his own survival techniques and morality.

"L'Inutile Beauté"

This is an extraordinary *nouvelle* by any standard. One might assume that it is being included for study here, in a chapter about the premises of Darwinism, since its title makes mention of beauty which, apart from its aesthetic sense, may be taken for a sexual lure. But we leave the dynamics of sexuality and reproduction for a later chapter. Instead, we look now at "L'Inutile Beauté" for another reason having to do with the thematics of a remarkable transcribed conversation between two characters.

The text is divided into four numbered sections or chapters. The first, second and fourth of these relate the story of Gabrielle, the beautiful Comtesse de Mascaret, whose husband's jealousy has reduced her to a life of pregnancy and domesticated motherhood; in order to break the grip of the Count's tyranny on her, Gabrielle tells him that one of their children is not his, and refuses to tell him more. After six years of separate lives during which the Count philanders on his own, she tells him that it was all a lie, and they end up as friends.

It is the third section of the text which is of interest to us at this point, however. This part of the *nouvelle* is set in an evening at the Opera, during an intermission, when two young men, chatting on the floor of the theater and surveying the balconies, spot the beautiful Comtesse de Mascaret and begin discussing her life and marriage. One of them, Roger de Salins, launches into a disquisition on nature, life and ontology that relates very much to the premises of Darwinism.

The point of departure is Roger's view of Gabrielle's eleven years of forced, cloistered motherhood. Roger laments "...cette abominable loi de la reproduction qui fait de la femme normale une simple machine à pondre des êtres" (CN2, 1216). This "law" is, of course, a product of culture and not of nature, for the most part, but Roger's friend, Bernard Grandin, replies, "Que veux-tu? c'est la nature!" as if to say that the tyranny of husbands is a law of nature. Roger ignores the fact of the dichotomy and goes on in a lengthy soliloquy on nature properly understood and humanity's place in it.

He states that nature is our enemy and that we must always struggle against it, "...car elle nous ramène sans cesse à l'animal" (CN2, 1216). At this point Roger sets God in opposition to humankind as creators: by Roger's account, it is not God who created what is beautiful or ideal on earth, but the human mind; God, he says, created only

> ...des êtres grossiers, pleins de germes des maladies, qui, après quelques années d'épanouissement bestial, vieillissent dans les infirmités, avec toutes les laideurs et toutes les impuissances de la décrépitude humaine. Il ne les a faits, semble-t-il, que pour se reproduire salement et pour mourir ensuite, ainsi que les insectes éphémères des soirs d'été. J'ai dit "pour se reproduire salement"; j'insiste. (CN2, 1216)

Roger then argues that a more thoughtful or considerate God would have made the reproductive process differently, so that it would not be a base function. So far this is still a question of nature set in opposition to culture.

It is at this point that Roger's remarks veer off into a consideration of the origins of life. God, to him, is a "monstrueux organe créateur...qui sème par l'espace des milliards de mondes" (CN2, 1217). From here, Roger's view of the universe is a mechanistic, materialistic one, and that universe is governed by principles of chance:

> Il crée parce que c'est sa fonction de Dieu; mais il est ignorant de ce qu'il fait, stupidement prolifique, inconscient des combinaisons de toutes sortes produites par ses germes éparpillés. La pensée humaine est un heureux petit accident des hasards de ses fécondations, un accident local, passager, imprévu, condamné à disparaître avec la terre, et à recommencer peut-être ici ou ailleurs, pareil ou différent, avec les nouvelles combinaisons des éternells recommencements. Nous lui devons, à ce petit accident de l'intelligence, d'être très mal en ce monde qui n'est pas fait pour nous, qui n'avait pas été préparé pour recevoir, loger, nourir et contenter des êtres pensants.... (CN2, 1217–1218)

Darwin himself could not have phrased his own premises better. With passing, token allowance for a divine creative force, the universe is portrayed as an ambiance of disorder where the only laws of further creation are the laws of chance and combination. In the middle of this material world and universe, humanity (with its faculty of reason) is an accidental occurrence, destined for eventual extinction, and in the meantime not particularly at home in a hostile environment where life is a struggle.

Human thought, the traditional saving grace of humanity in Christian ontology, is for Roger

> Une fonction fortuite des centres nerveux de notre cerveau, pareille aux actions chimiques imprévues dues à des mélanges nouveaux, pareille aussi à une production d'électricité, créée par des frottements ou des voisinages inattendus, à tous les phénomènes enfin engendrés par les fermentations infinies et fécondes de la matière qui vit. (CN2, 1218)

This is exactly the materialism that Darwin posits in offering his theory of life and evolution, and which Darwinian thought since Darwin has gone to great lengths to justify.

Roger insists that a providential God would have made a different world to receive a human being intentionally endowed with reason, not this

> ...petit parc à bestioles...où votre Providence imprévoyante nous a destinés à vivre nus...nourris de la chair massacrée des animaux, nos frères.... (CN2, 1218)

Herein is included the notion of the relatedness of earth's life species, where the animals are our brothers, and all species share the same circumstances of life in this world. Nothing sets this earth apart as especially the domain of humanity. Speaking of the earth, Roger asks,

> N'est-elle pas visiblement et uniquement disposée, plantée et boisée pour des animaux? Qu'y a-t-il pour nous? (CN2, 1218)

And of people? "Ceux-là seuls qui se rapprochent de la brute sont contents et satisfaits" (CN2, 1219). Humankind has had to adapt to foods and conditions suited better to other animals, and this includes the struggle among species who hunt each other according to their instincts; by way of explaining these circumstances, Roger asserts that

> Dieu n'a jamais prévu la douceur et les moeurs pacifiques; il n'a prévu que la mort des êtres acharnés à se détruire et à se dévorer. (CN2, 1219)

Finally Roger brings his argument back to the opposition of nature and culture: in the inhospitable environment detailed above, humanity has had to exercise its industry and imagination to render its world a little more habitable—all this "malgré la nature, contre la nature". And the more civilized and intelligent we are, "...plus nous devons vaincre et dompter l'instinct animal qui représente en nous la volonté de Dieu." (CN2, 1219)

Extinction

The issue of extinction is a crucial one for evolutionary theory, and certainly one that should be addressed in a Darwinian reading of any kind. On the "tree of life," a branch or twig eventually stops growing and subdividing; the metaphor is of course for the extinction of a genetic line. It is theorized that species eventually find themselves outmatched by the environment or by their enemies; unable to adapt and evolve a solution to prolong life, the species dies out. The reasons theorized are primarily two: catastrophe and failure to adapt.

Maupassant presents war as the sort of force which could bring about cataclysm sufficient to threaten the viability of life for the species, though the notion of species extinction is not broached directly, and the threat to life posed by the chaos of war is only an implication drawn from the many war settings in the stories. Likewise in "L'Homme de Mars" there is a hint that the cultural or technological superiority of interplanetary neighbor species could bring about the extinction of human life on earth (CN2, 1008), especially since the earth is portrayed as "une poussière invisible" and humankind "cet insecte imbécile" in the character's view (CN2, 1004).

More proximate is the motif of personal or family extinction, as exemplified in the failure or unwillingness to leave progeny. "La Reine Hortense" (CN1, 802–809) makes this issue a central motif: on her deathbed, Hortense has delusions of a past life with spouse and numerous children, but the images belie the solitude of her real life. The figure of the old aunt who regrets having no offspring recurs in *Une vie* (la tante Lison, R, 37 ff.) and in "Par un soir de printemps" (CN1, 309–313). The short story "Après" (CN2, 1246–1251) is the narrative of an old country priest who explains why he chose a life of celibacy, namely to avoid the sorrow of losing a loved one to death; to sidestep grief he chooses to end his genetic succession.

The matter of genetic succession is implicit in the motif of parentage to be discussed in chapter 6, though we might mention here in passing the concern of a character like "Monsieur Parent" (CN2, 580–617) for his lineage when he discovers that his only "son" was fathered by another man.

In this chapter we have seen a preponderance of evidence attesting to Maupassant's acceptance of the premises that serve as the foundations of Darwinian thought. Succinctly summarized, they are the relatedness of life species on a "bush" or "tree" model of representing their proximity and common ancestry, the principles of chance and materialism in a world of physical life dynamics, and a human ontology not perforce distinct from that of other animal species, but instead one in which humanity shares characteristics of morphology and behavior with animal species. The evidence has not been spotty or isolated and disparate. Rather it has shown a consistent world-view with indications distributed throughout the novels and short stories. With the premises established, then, we can move on to the evidence of Darwinian theory proper in the fiction of Maupassant.

NOTES

1. Forestier in Maupassant, CN2, 1360, note 3.
2. Naomi Schor makes lucid commentary on the blurring of lines between animal and human in reference to what she calls the "naturalization of woman" in realist fiction. See Schor, *Breaking the Chain: Women, Theory and Realist Fiction* (New York: Columbia University Press, 1985) 130.

Chapter 5: Hostility, Amorality, Struggle and Selection

The centerpiece of Darwin's theory, what was his innovative contribution to science's understanding of life, is the notion of selection. Evolution happens, in "this view of life," only as an after-effect, an ensuing consequence, of selection. Species change when anatomical or behavioral adaptations become the norm, and this only after adapted individuals survive in the struggle for life long enough (or well enough, or in large enough numbers) to reproduce and pass genes along to new generations. Species change, or evolution, is thus a by-product of adaptation which is itself a survivor of selection. Positive, helpful adaptations are "selected" or favored by a blind, amoral, apathetic and inhospitable natural environment whose forces (limited resources, rigors of climate or terrain, etc.) tend to push the fittest to the top and the less fit to the bottom.

Social Darwinism takes this interpretation of life in nature as a metaphor for life in human society. The same mechanism of selection, a function of natural forces working upon individuals engaged in a struggle for survival, is superimposed on the life of the human species. Substitute society for the natural environment, and social forces (economics, sociological pressures, demographics, etc,) for the pressures of life in the wild, and you arrive at a Darwinist model for human society referred to generally as Social Darwinism. The latter is customarily taken to be a metaphorical application of principle, but in different hands it has differing degrees of correlation with the tenor of the metaphor.

Motifs of these aspects of Darwinism recur in the novels and short stories of Maupassant, and it is to these that we turn our attention in this chapter. For the most part, they relate to human characters and their lot in this corpus of fiction, but to some extent they constitute a commentary on the world in which all of life makes its way. Maupassant's inclusion of the notion of the hostile universe, its order of amorality, and the contrasting principles of cruelty and altruism (or cooperation) will be our first concern. From there we will proceed to a study of competition, the struggle for life, natural selection and the survival of the fittest in the novels and short stories. Lastly we will consider motifs of Social Darwinism and the planned arrangement of parentage known in its extreme form as eugenics.

The hostile universe

It has already been pointed out that Maupassant was not one to fall in lock step with movements or ideologies for the mere sake of it, but it is also true that he is in concert with the other French Naturalists where the matter of the hostility of universe is concerned. As a group they portray the world as an inhospitable and unforgiving environment for human endeavor, and Maupassant adopts that *Weltanschauung* almost without exception in his fiction. Passages set in Corsica, North Africa or the wilds of Auvergne all unfailingly depict a landscape unfavorable to human habitation, and even animal species (e.g. the wolves in "Le Loup" CN1, 626, or the dog in "L'Auberge" CN2, 794–795) find adversity in the environment. Rarely in Maupassant is the concept of God brought into play, either as responsible for hostile circumstances or as the savior from those conditions; occasionally clergymen or –women appear to preach ecclesiastical morality or metaphysics, but this does not come as some *deus ex machina* to save characters (or even offer them much comfort) in the face of the world's hardships.[1]

It is the "impitoyable nature" which Georges Duroy laments (R, 395) in what is, for him, a rare moment of concern for the oppressed social castes, perhaps because he himself rose out of their ranks. Neither nature nor society pauses to tend to their needs, and as for heaven, the narrator recounts Duroy's thoughts as he waits in a church for a lover's rendezvous:

> Puis il réfléchit que ces gueux croyaient au moins qu'on s'occupait d'eux là-haut et que leur état civil se trouvait inscrit sur les registres du ciel avec la balance de la dette et de l'avoir. "Là-haut.—Où donc?" (R, 395)

There are no scales of justice for debts and possessions, and there is no heaven to redress wrongs. There is only nature and the world as they exist, and this is certainly why immediately Duroy "...jugeant d'une pensée la création, prononça, du bout des lèvres: 'Comme c'est bête tout ça.'" (R, 395).

On rare occasions in Maupassant, nature shows a slightly less than hostile face, as if to say that at some moments it presents favorable circumstances for the fittest. Chapter 2 of *Une vie* shows a young Jeanne before her bad marriage, very much in harmony with nature, the favorable nurturer (R, 16–19); that picture soon changes after her honeymoon as she returns from Corsica to the autumn rains of Normandy; she wonders,

> Etaient-ce la même campagne, la même herbe, les mêmes arbres qu'au mois de mai? Qu'étaient donc devenues la gaieté ensoleillée des feuilles, et la poésie verte du gazon où flambaient les pissenlits...? (R, 63)

Maupassant's nature is never prodigal with its favor: seasons change, people age, and the world turns. Jeanne's life becomes an ordeal, prompting her to think,

> Tout n'était donc que misère, chagrin, malheur et mort. Tout trompait, tout mentait, tout faisait souffrir et pleurer. Où trouver un peu de repos et de joie? Dans une autre existence sans doute! (R, 124)

It would have to be in another existence, for this one affords little peace and joy. The protagonist's "life" in *Une vie* is emblematic for all of Maupassant: any favor in nature is brief, and the hallmarks of her experience are disillusionment, disappointment and the deaths of those close to her.

When nature is kind, it is by omission, as in the case of "Ça ira" (CN2, 572–579), the character whose nickname gives the title of the short story. In this text, the narrator comes by chance upon an old flame, now the keeper of a tobacco shop in Barviller, where the narrator happens to get off a train; as they catch up on old times, she tells him of her hard knocks, but also of her chance meeting with a generous man who ended up the father of her child, and of the good turn her fortune took when her benefactor was able to land her a tobacco shop license. She is now self-sufficient and more or less content with life. Chance more than anything else occasionally provides shelter from nature's customary cruelty.

Other than that, it is said as a generality in "Le Vagabond" that "...la nature, la grande mère aveugle, est inéquitable, féroce et perfide" (CN2, 858), and this is the face of nature that Maupassant shows for the most part. Nature is responsible for the central motif in "Adieu" (CN1, 1246–1250) as it is in *Fort comme la mort*, namely the ravages of time on youth and beauty in people. Therre is little to mitigate the rigors of the human condition, as the narrator of "Solitude" (CN1, 1255–1259) broods, "...notre grand tourment dans l'existence vient de ce que nous sommes éternellement seuls"; even those who are in love

> ...cherchent, comme nous, comme toutes les créatures, à faire cesser leur isolement, rien que pendant une minute au moins; mais ils demeurent, ils demeureront toujours seuls; et nous aussi. (CN1, 1255)

Not a particularly sanguine view of life or the of the world. That world is, in this novelistic vision, an inhospitable place for human life and a hostile environment for the survival of any life form. The narrator of "Blanc et bleu" (CN2, 448– 452) phrases the thought most concisely:

> ...je pensais à cette pauvre et petite humanité, à cette poussière de vie, si menue et tourmentée, qui grouillait sur ce grain de sable perdu dans la poussière des mondes, à ce misérable troupeau d'hommes, décimé par les maladies, écrasé par les avalanches, secoué par les tremblements de terre, à ces pauvres petits êtres invisibles d'un kilomètre, et si fous, si vaniteux, si querelleurs, qui s'entretuent, n'ayant que quelques jours à vivre. Je comparais les moucherons qui vivent quelques heures aux bêtes qui vivent une saison, aux hommes qui vivent quelques ans, aux univers qui vivent quelques siècles. (CN2, 450–451)

Certainly human existence is contingent, but the narrator here is also pointing up the fact that the same holds for all species of life on earth, and furthermore that the earth itself is a material environment, subject to laws of physics and subject to eventual decline. Absent entirely is the Romantic notion that the earth is a protector, nurturing flora and fauna in idyllic harmony; in its place is the Darwinian view of a world where life arises as an accident and must struggle at every moment for survival.

When the characters' lives are circumscribed by war, of course, the hostility of the environment is graphic. Maupassant situated a number of his *nouvelles*, certainly some of his best known, in the Franco-Prussian war of 1870–71. One of the most engaging qualities of "Mademoiselle Fifi"(CN1, 385–397) or "Boule de suif" (CN1, 83–121) is the atmosphere of impending cataclysm and the frailty of human life in the storm of war. In "Deux amis" (CN1, 732–738) the chaos and danger of war is set in stark contrast with the friendship of two old fishing partners, now in uniform, who run across each other during the war. They decide to put the war aside and go to their old fishing spot for a day, when they are captured by Prussians; refusing to divulge their password, they are executed. In this and numerous other *nouvelles*, the war itself becomes a force of nature in that it dictates action as well as the environment in which characters find themselves engaged in a struggle for survival.

In the preceding chapter we had occasion to discuss the story "L'Inutile beauté" at some length, and we need not reiterate what was said there. But in the present context of the hostility of the universe, it is useful to highlight the gist of Roger de Salins' remarks concerning nature and the human condition therein. He says, "...je dis que la nature est notre ennemie, qu'il faut toujours lutter contre la nature, ...". The life that goes on within it is, in his view, spent in "quelques années d'épanouissement bestial," and against the antagonistic forces of nature, human weakness and decrepitude are no match (CN2, 1216).

There is an aspect of Maupassant's literature of the fantastic that is pertinent to the depiction of the hostile universe. Even though it relates to a thematic of the supernatural, it betokens a view of the universe as inimical to peace and order. It is the motif of the double, the thematic of a spirit which enters either the character's living space or the character himself. Paul Lefebvre sees in this motif a direct link with Darwinian thought, saying of the end of human domination of the earth (in relation to "Le Horla" particularly) that the text represents one of the first meditations on the uncertain future of the human race from a Darwinist perspective.[2] This interpretation is warranted by the effect of the double on the world of the human as well as by the depiction of nature as an environment which is entirely beyond human control.

The motif of a doubling or possessing spirit figures in no fewer than twelve of Maupassant's works of fiction (counting both versions of "Le Horla").[3] Clearly it is more than a passing literary fancy in his writing. The double (along with the accompanying effect of insanity, in many of its appearances) changes all the assumed rules about life on earth for characters who encounter it. Existence is no longer comprehensible, on the one hand, and the double deprives the character of peace and sanity, rendering his life a fearful experience with no hope of escape or aid. Thus the double adds an element of hostility and confusion into an environment already bereft, as we have seen, of order, benevolence and divine intervention. Even if the thematic here in question is supernatural rather than grounded in the natural world that is the exclusive purview of Darwinism, it is here a matter of the world set before the reader in the universe of fiction, and this universe is one where logic, order, progress and teleology are luxuries not afforded—and this is the point—just as in the Darwinist view of life.

Amorality in nature

To the same extent that the Darwinian world is without design or purpose, it is also amoral. It is one of the harder aspects of Darwin's theory with which to come to grips for a reader accustomed to Hollywood endings, and it is one which gave concern even to Darwin himself as he pondered to sense of predatory habits of exploitation that seem, at first glance, to be cruel and downright nasty. Darwin's reply has remained, that human morality is a code of values to which the natural world does not answer: nature knows no right or wrong, it knows only survival in circumstances that are most often adverse.

Biologist Richard Dawkins defines the Darwinian perspective succinctly in his discussion of the much-discussed case of the ichneumon wasp which is known to paralyze a caterpillar before depositing its eggs in the host; the larva uses the host for nourishment, carefully keeping it alive for its own needs and without any measure of concern for the caterpillar's comfort:

> It is better for the genes of Darwin's ichneumon wasp that the caterpillar should be alive, and therefore fresh, when it is eaten, no matter what the cost in suffering. Genes don't care about suffering, because they don't care about anything.
>
> If Nature were kind, she would at least make the minor concession of anesthetizing caterpillars before they are eaten alive from within. But Nature is neither kind nor unkind. She is neither against suffering nor for it. Nature is not interested one way or the other in suffering, unless it affects the survival of DNA. (131)

It is rare for a primary narrator in Maupassant's fiction (or even a secondary narrator, for that matter) to make an explicit comment on the rectitude of character's actions or of conditions in which action unfolds. Judgment may be implicit in the way things are presented or between the lines of what is said in narrative discourse, but narrators tend almost always to forgo moralizing as if to accept the world on its terms of indifference to moral values. The notable exception to this general principle is the physician-narrator who comments on lamentable circumstances from a medical perspective of one professionally engaged in a battle with nature's destructive tendencies. "Le Baptême" (CN2, 436–441) gives an example of such a commentary (Donaldson-Evans, 1994, 97–110). For the rest, the world is taken as it is, in recognition of the fact that moral values cannot be imposed on it with any effect.

Perhaps the most succinct assessment of the amorality of the universe comes at the very end of *Une vie*. After Jeanne has come full circle from tranquility, through strife and near-deprivation, and back to a state of relative calm and well-being, the narrative concludes with her remark, "La vie, voyez-vous, ça n'est jamais si bon ni si mauvais qu'on croit." (R, 194) The moral status quo of the universe is somewhere between good and evil, neither one nor the other, that is to say, amoral. And this is the same Jeanne who concluded long before in the novel that "...l'être moral de chacun de nous reste éternellement seul par la vie" (R, 54). To her disappointment she discovers that there are no moral standards that bind everyone and everything.

There are two noteworthy cases in the short stories where a central character's lack of moral fiber allows him to survive and prosper in his pursuits. In "Saint Antoine" (CN1, 772–779) the protagonist allows someone else to be arrested and executed for the murder of a Prussian soldier which he committed; in the end Antoine (nicknamed ironically "Saint-Antoine" for his *bon vivant* lifestyle) shows no remorse or sense of guilt for the way things turn out in his favor. Elsewhere, in "Le Mal d'André" (CN1, 912–917), a captain Sommerive is content to let a nursemaid take the blame (at the expense of her job) for what he did to his mistress' baby son: in order to have the mistress to himself and to frighten the baby into fearing his mother's bed, the captain repeatedly pinches him when his mother brings him to bed with her. When the woman's husband returns after a week's absence, the baby is seen to have numerous black and blue marks; the nursemaid is accused of abusing the child and is fired; the text ends with the observation that the nursemaid was prevented from finding another job on account of "her conduct," while captain Sommerive has had his way with impunity. In a world where it may be said, as we often say, "there is no justice," the point is that there is no effective moral agency to see to the triumph of good or even of order.

The breakdown of morality in the making of human choices is, in Maupassant, a reflection of the amorality of the universe in which characters live. In "En mer" (CN1, 739–744) the crew of a fishing ship is buffeted by a raging storm that threatens to wreck the vessel; this is an easily understood metaphor for the contingency of life in a dangerous world—a ship cast about by wind and waves on a stormy sea.

After the storm calms somewhat, the still rough sea causes the younger brother of the ship's captain Javel to slip and get his arm caught between the fishnet and the rail; after some thought, the captain makes the decision to sacrifice his brother's arm rather than incur the expense of cutting the trawl to free the trapped limb. The sea evidently "cares" as much for the well-being of the trawler and its crew about as much as the elder Javel, having considered his options, cares for his brother's arm. Captain Javel makes his choice coldly and dispassionately, by a standard that sets aside the concept of morality. The same could be said of nature personified in the fiction of Maupassant.

Other reflections of nature in human characters' amorality are visible in the short stories. In "Un fou" (CN2, 540–547) we encounter the journal of a judge who has died after a career in the highest esteem of his community. The journal reveals the judge's fascination with killing:

> Pourquoi donc est-ce un crime de tuer? oui, pourquoi? C'est, au contraire, la loi de la nature. Tout être a pour mission de tuer: il tue pour vivre et il tue pour tuer.—Tuer est dans notre tempérament; il faut tuer! (CN2, 541)

The journal entry goes on at some length to portray killing—by animals in the wild and by humans in war and in justice—as an integral part of life. Then the journal reveals that the judge has given rein to his impulses by killing two people; subsequently he let investigations go on and presided over the trial of the second victim's nephew who was found guilty and executed by the judge's order. This madman takes his cue from the nature which he observes around him, nature which has fostered an order of predation and which brings about the death of all living organisms. The judge's dementia is his failure to recognize that morality establishes rules of conduct for human society, but even for its bizarreness, the judge's insanity points up the amorality of the natural "order" that is reflected in his actions.

Maupassant's numerous fictional representations of insanity, and even his recurring use of the motif of the double, betoken a world without reason, without order or morality or compassion. We have already discussed some texts whose thematics fall in this category: "Un cas de divorce" (CN2, 777–783) and "La Petite Roque" (CN2, 618–650); these two texts, like "Un fou," involve the proceedings of the justice system, a fact which Maupassant uses to point up the stark contrast between the utter amorality of nature and the attempts society makes to impose moral order on a world that is indifferent to morality.

In thematically related cases, with "La Ficelle" in chapter 3 above and with "Le Vagabond" as well, we noted that society can use the wheels of justice and social censure to grind innocent individuals. In "Le Vagabond," it is a case of what Fredric Jameson sees as an underlying message to the poor and disenfranchised, to stay in the place that society has cut out for them (187–189). In other instances, as in "La Ficelle," it appears that society's motivation is to avoid being duped, even at the cost of prosecuting those who are less than guilty. In the short stories, there are also cases of the inverse situation where society is duped by the guilty, that is to say, where the moral system is short circuited by those who would profit by its ineffectiveness. We have seen the example of the judge in "Un fou" and the mayor Renardet in "La Petite Roque" who both fool society into believing that they are upstanding pillars of the community when, in fact, they are pathological murderers. In "L'Orphelin" (CN1, 848–854), the adopted son and heir of a severely disfigured woman is suspected of his stepmother's murder. The narrator details the young man's past conduct which caused his stepmother's abiding fear of him, and all of the people in the area assume he is guilty. But he has something of an alibi which is sufficient to derail his prosecution. In the years after the murder, he is able to gain society's confidence by dint of joviality, and he eventually is elected mayor of the *commune*. There is also the case of "Moiron" (CN2, 984–990), narrated by the former prosecutor who had prosecuted the defendant. Moiron is a school teacher who, after losing all three of his own children to illness, goes on a spree killing the children of others. Condemned by overwhelming evidence, Moiron lies his way to a commutation of the sentence of death by convincing some influential people that he was wrongly convicted. Finally, on his deathbed he confesses his guilt. These are all instances of a moral order circumvented and used against itself, to point up the inefficacy of morality in the universe as it is, devoid of moral order. Typically of Maupassant—who obviously was fascinated by the notion of the double and about whose novel *Pierre et Jean,* it may be said that everything in that novel is dual[4]—there are two mutually offsetting points on the human compass. Moiron on his deathbed phrases it thus, as he speaks of the God he once adored:

> ...ce bon Dieu—le Dieu qu'on nous enseigne à aimer, et pas le Dieu faux, le bourreau, le voleur, le meurtrier qui gouverne la terre. (CN2, 988)

As we saw with Jeanne of *Une vie*, whose final assessment was that good and evil cancel each other out in an amoral world, Moiron's two "Gods" annul each other, with the same effect of removing good and evil as meaningful terms in an amoral universe. As the "skeptical genius" is quoted in "La Légende du mont Saint-Michel," "Dieu a fait l'homme à son image, mais l'homme le lui a bien rendu" (CN1, 679)(and Louis Forestier links this quotation to Schopenhauer for us[5]), attesting to Maupassant's general view of theology as anthropomorphic. The narrator of this legend goes on:

> Ce mot est d'une éternelle vérité et il serait fort curieux de faire dans chaque continent l'histoire de la divinité locale, ainsi que l'histoire des saints patrons dans chacune de nos provinces. Le nègre a ses idoles féroces, mangeuses d'hommes; le mahométan polygame peuple son paradis de femmes; les Grecs, en gens pratiques, avaient divinisé toutes les passions. (CN1, 680)

The clear implication to be drawn is that God, as a human artifact, is absent as a force of morality or as an agent of justice from the events of the world. This is not to say that, in Maupassant's fictional vision, the world and the universe are evil; rather it is to say that they are neither evil nor inherently directed towards the good, but amoral, unconcerned and unaffected by moral principles of any kind.[6]

Cruelty, altruism and cooperation

Given what has gone before concerning the hostility and the amorality of the environment, it is understandable that one would be as likely to come upon examples of apparent cruelty as examples of the opposite. With no moral force policing the world, there is no particular disincentive to cruelty, either gratuitous cruelty or that which is the side-effect of self-interest. Conversely, self-interest can produce acts of altruism or cooperation, and altruism can also appear in "random acts of kindness" as well. As one might expect in the fiction of so pessimistic an author as Maupassant, there tend to be more instances of cruelty than altruism in the novels and short stories.

We begin with a recurring motif in the corpus, that of cruelty to animals. For the most part, this does not occur out of humans' self-interest (and we exclude from this motif the numerous cases of exploitation of animals through the hunt which crop up in Maupassant). Above mentioned was the donkey worked to death in *Mont-Oriol*; it is assumed by the protagonist Christiane that the animal had belonged to a desperately poor family for whom the animal's death would cause additional hardship. Then there is the case of "Pierrot" (CN1, 570–575), the dog taken in by two older Norman women who decide that they do not wish to suffer the expense of keeping a dog. They then abandon the animal in a pit outside of town reserved for that function. At first they continue to feed it out of guilt, but then they rationalize that they may be feeding other animals inadvertently, so they abandon it entirely. In "Pierrot" it is untrammeled self-interest (in the vehicle of Maupassant's habitual criticism of Norman avarice) which prompts the choices of the human characters.

A kind of self-interest enters into the actions of the fanatical Abbé Tolbiac in *Une vie,* and curiously it is his puritanical morality that leads him to his cruelty. When he comes upon a group of children watching a dog give birth (of which the narrator observes, "C'était un jeu pour eux, un jeu naturel où rien d'impur n'entrait." R, 141), he beats the dog and its pups to death with his umbrella. The identical episode recurs in the short story "Le Saut du berger" (CN1, 378–379).

The other instances of cruelty to animals are more gratuitous, whether they happen out of sport (as in "L'Ane" CN1, 902–911, where two marauders kill a donkey for target practice) or out of vengeful laziness (as in "Coco" CN1, 1149–1152, in which a boy who is charged with the care of an old horse causes the animal to starve to death rather than go to the trouble of taking it to pasture). In "Sur les chats" the narrator recounts a childhood memory of allowing a strangling cat to die, watching "avec une joie frémissante et cruelle", rather than extricating the animal (CN2, 693). These cases of cruelty clearly fly in the face of any sense of humaneness for animals, and set characters up as high-handed brutes with no regard for other species. Here humanity willingly steps in to serve as the agent of the hostile environment for other animals.

In the circumscribed environment of human society there is no shortage of instances of interpersonal cruelty. Cruelty can be a means of self-assertion, always a matter of interest in an interpretation whose model is the natural environment: Julien, the protagonist's callous husband in *Une vie* reacts to the general laughter at a servant's gaffe by slapping the servant and imposing fear in place of levity (R, 70–71). Cruelty can also be a sanction for conduct that is judged to be unacceptable. Here, of course, morality is in effect—at least to the extent that it constitutes a set of principles according to which the "environment" is ordered; when it comes to enforcement of that order, however, morality turns a blind eye to cruelty and the environment becomes a metaphor for the natural world. As we shall see, the resemblance is not lost on Maupassant who chooses images from the animal world to describe human relationships and social dynamics.

Young women are treated cruelly for becoming pregnant outside of marriage, or even for being the victim of rape. In "Histoire d'une fille de ferme" (CN1, 225–243) a young girl is abandoned when she becomes pregnant and is forced to give the baby up; subsequently she is forced by her employer to marry him. And there is Rosalie in *Une vie* who is seduced and abandoned by her employer, Jeanne's husband, and cast out of the household by the man when she delivers her baby. In "Madame Baptiste" the title character is given the nickname, having been raped by a certain Baptiste, and is ostracized by the community; even after her marriage to a decent man from outside the village, she is driven to suicide by a public insult, and in a final ill turn, she is denied a Catholic burial since she took her own life. Society, even in the institution of its religious morality, shows no compassion whatever in an environment which parallels a Darwinist view of nature.[7]

Mistreatment of the elderly is another facet of the motif. We have already looked at the cruelty shown the grandfather in "Une famille (CN2, 763–768). In both "Rencontre" (CN1, 440–444) and "Humble drame" (CN1, 1015–1020) there is the story narrated by an old woman of her abandonment by her family. "Misère humaine" (CN2, 752–757) depicts a feeble old woman left to fend for herself in the busy Parisian metropolis that is apathetic to her lot. "Le Vieux" (CN1, 1130–1137) gives a glimpse of the younger generation's heartless attitude towards the elderly: when a dying old man lingers, the surviving family makes the funeral arrangements a bit prematurely, and then resents the expense they incur as a result.

In a Darwinian environment (again, devoid of the aid afforded by a humane system of morality) as Maupassant portrays society to be, for the most part, the infirm are cast aside and left to tend to their own needs. "L'Aveugle" (CN1, 402–405) tells of the blind man abused by family and neighbors who are relieved at his eventual death. For the title character, the narrative gives the metaphor of the feeble chicken killed by the rest of the flock, and does so in rather Darwinian language:

> ...car aux champs les inutiles sont nuisibles, et les paysans feraient volontiers comme les poules qui tuent les infirmes d'entre elles. ...et il devint un souffre-douleur, une sorte de bouffon-martyr, de proie donnée à la férocité native, à la gaité sauvage des brutes qui l'entouraient. (CN1,

403)

The protagonist of "Le Gueux" (CN1, 1225–1230) is at 15 the victim of an accident that leaves his legs crushed. The surrounding villages soon tire of his begging, and in his desperation he tries to steal a chicken. He is beaten by the farm family and hauled off to jail where he dies of neglect. Again, the infirm individual is the outcast, the object of scorn, and is in effect put to death by the "herd" that has no use for him.

This depiction of the herd, held up for comparison to human society, is perhaps the starkest illustration of the Darwinian model written into the fiction of Maupassant. We saw something of the same phenomenon in our study of "La Ficelle" in chapter 3, where the human herd gangs up on a defenseless individual as if to enforce the principle that the weak should not survive. In nature, of course, Darwin saw a blind force which selects against the weak as a function of their inability to survive and hence to reproduce; in the harsher versions of Social Darwinism, this protocol is (wrongly, for any ethical person) superimposed on human society, ostensibly as a simple fact of life, but the comparison is oblivious to anything humane in the culture that can—or should—insulate it from nature. Maupassant subtly points up this lapsus, and typically does it only implicitly.

Simon in "Le Papa de Simon" is another character in whom the "herd" spots a weakness which it then persecutes. His weakness is that he has no father, unlike the other boys, and for this he is ridiculed and bullied. The narrator of the text calls up the same metaphor of the infirm chicken, and the other boys who consider Simon "un être hors de la nature" are described as animals in a truly Darwinian turn of phrase:

> ...et ces fils des champs, plus proches des bêtes, éprouvaient ce besoin cruel qui pousse les poules d'une basse-cour à achever l'une d'entre elles aussitôt qu'elle est blessée. (CN1, 75)

Other cases of human cruelty abound in the corpus. The *nouvelle* "Boule de suif" is one that is very well known among Maupassant's writings. Here a woman, understood by all to be a *courtisane*, is singled out by her traveling party to answer a Prussian officer's sexual demands. The group includes two austere nuns, several fine *bourgeois*, and even a cynical republican politico, but all of them tolerate the abuse of Boule de suif if it means they can get their freedom from the Prussian's arrest. In this text, war and the Prussian occupation create an environment of pressure that brings out the worst in ostensibly upstanding characters. Morality goes by the board when survival (or even convenience) is at stake.

There are two noteworthy cases in which a young man's thoughtless intervention in the lives of those around him cause harm to them. In "La Question du latin" (CN2, 797–805), a young man's Latin tutor loses his post because of the student's efforts to involve him romantically with a washerwoman. In "Mademoiselle Perle" (CN2, 669–684) the narrator one evening discovers his married host's lifelong love for the latter's adopted "sister" and decides to reveal it to the woman; in the aftermath of the havoc he causes, the young man simply grabs his hat and leaves, waxing disingenuously philosophical about how those whom he has hurt are really better off for his actions.

We conclude the survey of cruel conduct with a consideration of a short story, "Mon oncle Jules" (CN1, 931–938), interesting for its incorporation of the motif of kinship with that of cruelty. The secondary narrator, Joseph Davranche, tells his friend of his family's treatment of his uncle Jules when the latter, after going off to South America to make a fortune and pay the family back for his profligate years, returns poor to crush their dreams. The rest of the family disowns Jules in particularly heartless fashion; Joseph, a boy at the time, is moved to pity and cannot overlook the bond of kinship.

Let us pick up the thread of altruism where we left off, with Joseph Davranche. His charity towards his paternal uncle, which is rebuked by his hard-hearted mother, betokens a recognition of the genetic bond of kinship in the immediate circumstance. The text opens with the primary narrator's observation of his friend, Joseph, offering alms to a beggar; Joseph explains that his altruism was conditioned by his own family's lack of it. Where the support of kin is concerned, however, an interpretive explanation has been proffered by Darwinist thought, namely that it is a gesture toward favoring the individual's own genetic line.[8]

"Boule de suif" sets up a vivid contrast between the conduct of the traveling party and that of the central character herself. They show her cruelty for their own gain. She, for her part, first shares with them the provisions which she alone had the foresight to pack, but this generous gesture is a foretaste of the selflessness she will be called upon to show. In the end, she will sacrifice her dignity and her person for the benefit of the group. Her altruism assures the survival of the entire group.

There is a character who comes to the rescue of Simon in "Le Papa de Simon." A man who takes a romantic interest in Simon's mother decides to marry her and give Simon a father. On the one hand, his generosity could be seen to have a measure of self-interest: he cannot have the mate without adopting her son as part of the arrangement; on the other hand, his words and the implicit narratorial judgment lead the reader to believe that there is much charity in his act. At any rate, his gesture is protective, assuring Simon's survival in the society of his peers. He says to Simon,

> "Tu leur diras, à tes camarades, que ton papa, c'est Philippe Remy, le forgeron, et qu'il ira tirer les oreilles à tous ceux qui te feront du mal." (CN1, 82)

One plot serves as the line of action for three of Maupassant's short stories set in the context of the mayhem of the Franco-Prussian War. In these texts, "Le Mariage du lieutenant Laré" (CN1, 65–69), "Souvenir" (CN1, 362–366) and "Les Idées du colonel" (CN2, 162–167), a detachment of soldiers on a stormy winter night comes to the rescue of a young woman and her father, by unanimous decision to show charity to people in need. In one of these works, Lieuetenant Laré eventually marries the young woman, so there may be seen a measure of sexual motivation in the soldiers' empathy; the Colonel in another of the texts, concludes the narrative with the observation, "...nous aimons bien les femmes, nous autres Français!" (CN2, 167). Undeniably an act of altruism in all three short stories, the rescue of the young woman and her father may also offer the benefit of sexual (and thus, procreative) interest from a Darwinist perspective.

The altruism of the title character in "Clochette" (CN2, 851–855) is motivated by love and hence may be likened to the cases just considered. Here a young woman jumps from a high window to avoid discovery and save her lover from the compromise that would cost him his job, but in the process she shatters her leg and is left with a limp for the rest of her life. If we apply a Darwinist model of interpretation to this and other texts, we are constrained to separate the moral motivation from actions in order to see in them their effects on survival and differential success in passing genes along to a subsequent generation. Thus put, it sounds awfully heartless and unromantic. Let us make it clear, then, that we are studying only certain aspects of texts in this reading, and certainly not denying the existence or importance of other aspects; Clochette's act of self-sacrifice is laudable by any human standard, and her injury is pitiable. Our only interest in it at the moment is how it relates to the hostile environment and the struggle for survival. But we must note, too, that her self-abnegation, heroic in the word of the physician/secondary narrator who relates her story (CN2, 855), stands in sharp contrast with the cowardice and cruel abandonment of the young man she protects. His choices and their effects on Clochette point up the futility of love in the hostile environment of Maupassant's fictional world.

"Le Lit 29" (CN2, 174–185) is of interest at this juncture in that it relates to the motif of altruism in a directly biological way. In this narrative set in Rouen, the dashing Captain Epivent has been involved with the beautiful Irma for a year when the war breaks out and he is sent to the front. When he returns after the war, it is to learn that Irma is in the hospital, dying in the throes of syphilis. When he visits her in the hospital under the disapproving gaze of staff and patients, she informs him first that she had been raped by the enemy, and then that she had used her beauty to lure Prussians into sexual relationships and had refused medical treatment in order to spread the venereal disease among the enemy. Here we have the case of a character using a biological weapon and sacrificing her own health in order to maximize her effect. Unlike other heroines throughout the corpus (e.g. Clochette), Irma's motivation is not romantic, but patriotic instead. Her treatment at the hands of her beloved, however, is similar to that of others: she is abandoned to her fate by Epivent who is something of a cad, decorated though he is for his service to the country; Irma receives no such accolade for her self-sacrifice. She clearly recognizes her community as France and gives up her own well-being for the sake of the community's survival; the community, for its part, treats her with scorn and ostracism.

Not all altruism in Maupassant is greeted with disapproval, and there are at least two cases in which the act (or the intention) remains more or less the secret of the benefactor. Jean de Valnoix, the secondary narrator of "Le Père" (CN2, 975–979), shows compassion and generosity to an unwed woman in childbirth and receives an annual visit from her in her gratitude. It is interesting to note that his friend, the primary narrator, makes a point of contrasting Valnoix's earlier dissolute lifestyle with his presently productive and staid existence: the implicit message is that the latter is of more good to the community than the former.

Then there is the scene in *Pierre et Jean* in which Pierre is inspecting the ocean liner where he has accepted employment as ship's physician, and he passes through the steerage section where the very poor passengers are lodged. Clearly he is moved by the sight of their squalid conditions, and his instinctive judgment is that all hope is lost for these, the least fit to survive in the reality of the world. With "cette existence d'abominable misère" before his mind's eye,

> ...le docteur eut envie de leur crier: "Mais foutez-vous donc à l'eau avec vos femelles et vos petits!" Et son coeur fut tellement étreint par la pitié qu'il s'en alla, ne pouvant supporter leur vue. (R, 828–829)

Competition and the struggle for life

Obviously this is a motif central to a Darwinian model of interpretation. Competition is the demand made upon any life form in the natural environment, and the Social Darwinian perspective is quick to transpose the notion to the human social milieu. According to Diane M. Smith,

> The appropriation of the Darwinian struggle for existence had a distinct advantage for nauralist fiction; it provided an ideological framework for a more candid portrayal of the degradation and violence of working class life that bore the stamp of scientific validity. (76)

Given the character of the environment, both in nature and in Maupassant's fiction as we have thus far detailed it, survival necessitates competition in struggle. Whether adversity comes from the natural elements (as in "L'Auberge," CN2, 784–796 or "Le Mariage du lieutenant Laré, " CN1, 65–69) or from the prevailing socioeconomic conditions (as in *Bel-Ami*), the environment makes it incumbent upon the individual to struggle for survival, lest he or she perish. The ensuing struggle can be physical, financial, or even moral in instances (like that in "Boule de suif") where characters' principles come into play—this in glaring contrast to the amorality of the environment.

The struggle brings out all sorts of conduct in the fictional human characters, as desperate times call for desperate measures. "Idylle" (CN1, 1193–1197) shows a destitute, unemployed laborer aboard a train for Marseille who, not having eaten in two days, takes nourishment from a lactating woman whose breasts are engorged. The mores of culture and etiquette are trumped, and without much ado, by the necessities of survival. We have already examined the cases of "Le Gueux" (CN1, 1225–1230) and "Le Vagabond" (CN2, 856–867) in which both protagonists, decent enough at the outset, are reduced to criminal acts in order to survive. The imperatives of the struggle lead both of these characters to set aside the rules of living in human society; the Vagabond especially goes to some length in his thought process to justify his illicit acts—even those not directly related to his effort to survive—by holding them up to the injustice and heartlessness of the environment.

The struggle for survival is one engaged in most seriously, and the tenor of the texts mentioned is correspondingly somber. But the struggle can also be farcical in another setting. Even when, in the beginning of "Le Trou" (CN2, 831–837), the reader learns that the context is a murder trial, the action turns out to be farcical: two fishermen are fighting over precedence at a fishing hole, and someone is knocked overboard in the ensuing battle of fists and words.

Of course, the context of war in many of Maupassant's short stories serves to focus attention on the struggle for survival in a ferocious environment. The war itself is a crucible, a microcosm of the natural environment which tends inexorably toward the destruction of the individual. The ferocity of the circumstances brings out the worst, for the most part, in the characters caught up in them, and they devise various tactics for survival. The title character in "L'Aventure de Walter Schnaffs" (CN1, 793–801), for example, crafts a strategy of ready surrender in order to stay alive. In all the war texts, the war is the ultimate struggle for survival, a fierce and wild environment without quarter for anyone. It creates extreme pressures on characters, forcing adaptation of behavior for survival's sake. In "Les Rois" (CN2, 882–894), characters desperate for relief from the perils of battle stage an Epiphany celebration which becomes bizarre in the circumstances; and it is ended abruptly when a sentry shoots a deaf man who does not observe the sentry's challenge.

In more normal social settings, the struggle takes on purely Social Darwinian dimensions as it involves the struggle to survive economically in an environment defined principally by its economics. A city setting typical of Maupassant is encapsulated in the observations of Duroy on his lodgings early in *Bel-Ami*:

> Cela sentait la misère honteuse, la misère en garni de Paris. Et une exaspération le souleva contre la pauvreté de sa vie. Il se dit qu'il fallait sortir de là, tout de suite, qu'il fallait en finir dès le lendemain avec cette existence besogneuse. (R, 224)

The possible outcomes of this struggle are limited in number: either a truly Darwinian event of random chance can change the character's lot (as occurs in chapter 1 of *Bel-Ami* where Duroy meets Forestier on the street), or the application of unbridled ambition advances him or her (as is the case, again, in *Bel-Ami*) in a more Lamarckian order of evolution by acquired characteristic; finally, the character can be crushed by the environment and pass unnoticed in fulfillment of the predictions of Malthus for the least fit to survive in the environment of an uncaring economy. In any event, the textual actions in urban settings differ little from those in rural settings; in the latter, there is more intervention by the natural environment (and thematic juxtaposition with animal species, as we have seen), while in the urban tales the environment dynamics are socioeconomic for the most part; the outcomes are usually the same in both settings, for the misfortune of the characters. In both arenas, the environment favors ambition, either the heartless variety of Duroy in *Bel-Ami* or the conniving sort typical of the Norman peasants that populate Maupassant's stories (like "Le Petit Fût," CN2, 77–82, one example among many).

The well-known short story "La Parure" (CN1, 1198–1206) centers around the struggle to survive what is a financial disaster, brought about by misunderstanding and sheer bad luck. The young and beautiful Mathilde Loisel, disenchanted with the mediocrity of her social standing, gets an invitation to a society ball and borrows a necklace from a wealthy friend to complement her gown. Upon arriving home after the ball, she and her husband are shocked to find that the necklace has been lost. Assuming that its diamonds were real, they purchase a replacement to return to the friend, and set about paying the enormous debt. Mathilde loses her youth and beauty working menial jobs for ten years.

> Mme Loisel connut la vie horrible des nécessiteux. ... Elle connut les gros travaux du ménage, les odieuses besognes de la cuisine. Elle lava la vaisselle, usant ses ongles roses sur les poteries grasses et le fond des casseroles. Elle savonna le linge sale, les chemises et les torchons…; elle descendit à la rue, chaque matin, les ordures, et monta l'eau, s'arrêtant à chaque étage pour souffler. ... Mme Loisel semblait vieille, maintenant. Elle était devenue la femme forte, et dure, et rude, des ménages pauvres. (CN1, 1204–1205)

Losing the middle-class insulation from poverty she had, Mathilde is cast upon the environment to survive by the work of her hands. The struggle for financial survival (detailed by the narrator: "Il fallait chaque mois payer des billets, en renouveler d'autres, obtenir du temps" and so on, CN1, 1205) is harsh and takes a great toll on Mathilde physically. According to the narrator's description, she turns into a product of the harsh environment, losing the "jolie et charmante" aspect recounted in the text's opening sentence, and taking on the appearance and behavior of a hag as a result of the toil that need imposes upon her. Needless to say, the need and the environment in which it is acquitted are of human (and not natural) creation, socioeconomic in kind, but as severe as any imposing upon animals in the wild—this is the fundamental assumption of Social Darwinism, that the social milieu with its economic realities is a crucible in which people must compete for resources or fall by the wayside. The injustice and the uncaring severity of the environment is underscored in "La Parure" by he fact, revealed at the end in a chance meeting of Mathilde and her wealthy friend, that the original lost necklace was of costume jewelry. This is a perfectly coherent turn of events in a Darwinian reading: the environment, blind to matters of fairness and reason, is quite capable of bringing this fate on the individual as an effect of sheer chance and the adversity that is the order of all natural (and, by extension, social) existence. Chance and adversity are two of the *forces motrices* of the Darwinian view of life.

Another example of the literary character buffeted by chance and the economic facts of life in society is Pierre of *Pierre et Jean*. This individual awakens from a life of complacency to discover that, by genetic chance, his (half-) brother Jean stands to inherit the fortune of their mother's wealthy ex-lover, Jean's biological father. Pierre, in the meantime, is left to deal with the social implications of the mother's infidelity and with his own economic need. At first, Pierre decides to ply some ambition and make his own fortune in the practice of medicine:

> Le docteur se réveilla le lendemain avec la résolution bien arrêtée de faire fortune. ... Combien de médecins étaient devenus millionnaires en peu de temps! ... Et il serait plus riche que son frère, plus riche et célèbre, et content de lui-même, car il ne devrait sa fortune qu'à lui.... (R, 742–743)

These observations in the narrated mental process of Pierre are revealing for our purposes. Pierre is the loser in the game of genetic chance here, since Jean has the luck to inherit a fortune out of the blue. His first reaction is to compensate for his bad luck by competing, and he resolves to succeed so that he might have the satisfaction of creating his own destiny and overcoming adversity on his own. In other words, he would thus show himself fit for the environment and would assure his own survival.

That day at the noon meal with his family, Pierre expounds on his belief that the fit will survive on their own, while the merely lucky never have to prove their mettle:

> Pierre, à son tour, donna son avis. En somme, ce n'était pas la fortune qui faisait la valeur morale, la valeur intellectuelle d'un homme. Pour les médiocres elle n'était qu'une cause d'abaissement, tandis qu'elle mettait au contraire un levier puissant aux mains des forts. Ils étaient rares d'ailleurs, ceux-là. (R, 744)

Pierre's views are essentially Darwinian, that is, that the strong survive by the application of their natural gift to excel and outstrip the mediocre in the competition for resources. In comparison to the average individuals, the better endowed are numerically few and are rendered exceptional by their differential advantage. In this case, the advantage in question is intelligence, and with the use of the term "valeur morale," we can understand implicitly that Pierre also includes the virtue of ambition in his inventory of attributes for survival.

But in this novel, as we have mentioned, everything is dual. Pierre is given to opposites within himself. On the same afternoon, his resolution to compete runs out of steam:

> Comme il n'était pas encore quatre heures, et qu'il n'avait rien à faire, absolument rien, il alla s'asseoir dans le Jardin public; et il demeura longtemps sur son banc, sans idées, les yeux à terre, accablé par une lassitude qui devenait de la détresse. (R, 746)

He ends up taking a low-end job as a ship's doctor, very much demoralized by his situation and disinclined to compete for financial success.

Diane M. Smith makes the point about Naturalist fiction in general that

> The inevitability of the moral corruption of the young, particularly young females, is frequently expressed in naturalist fiction in terms borrowed from natural science that are couched in the language of the moralist. (78)

One response to the adversity of the socioeconomic milieu, a response that crops up fairly often as a motif in the fiction of Maupassant, is the recourse to prostitution. Women resort to this desperate measure either because there is no other way for them to survive or because it offers a lone way for a woman to climb socially.

An example of the former (from among many in the short stories) is detailed in "L'Odysée d'une fille" (CN1, 997–1003) where a young woman relates to the narrator her tale of how the hostile world drove her to prostitution in the streets. In this character's case, the institutions of society turned against her: an orphan, she is assaulted by her churchgoing employer whose jealousy forces her to leave her town; on the road she is raped by two gendarmes and subsequently abused by a judge; these and other circumstances leave her homeless and with no hope for decent employment. Her last remarks to the narrator are:

> Oh! ici, monsieur, c'est dur de vivre. On ne mange pas tous les jours, allez. Y en a trop. Enfin, tant pis, chacun sa peine, n'est-ce pas? (CN1, 1003)

The Darwinian dynamics of environment are here obvious, the difficulty of the struggle for life, the paucity of food sources, the reality of adversity for all.

An example of the recourse to prostitution as a means of material advancement, really only a difference of degree from the preceding sort, figures among other places[9] in "Yvette" (CN2, 234–307). The mother of the title character has, for some years, accumulated a measure of wealth as "une femme d'amour." When Yvette realizes the truth about her mother's past and present, she confronts her with the demand that they live as "honnêtes femmes." The mother defends her choices:

> Eh bien! oui , je suis une courtisane. Après? Si je n'étais pas une courtisane, moi, tu serais aujourd'hui une cuisinière, toi, comme j'étais autrefois, et tu ferais des journées de trente sous, et tu laverais la vaisselle, et ta maîtresse t'enverrait à la boucherie, entends-tu, et elle te ficherait à la porte si tu flânais, tandis que tu flânes toute la journée parce que je suis une courtisane. Quand on n'est rien qu'une bonne, une pauvre fille avec cinquante francs d'économies, il faut savoir se tirer d'affaire, si on ne veut pas crever dans la peau d'une meurt-de-faim; et il n'y a pas deux moyens pour nous.... (CN2, 287)

She goes on to say that the option of earning wealth at the stock market is not open to a servant girl, and that "Nous n'avons rien que notre corps, rien que notre corps." Such is the struggle for survival as Maupassant sees it, and he does not hide any of its unseemliness or brutality.

Darwin devotes the entire third chapter of his *Origin of Species* to the topic of the struggle for existence, and he concludes it thus:

> All that we can do, is to keep steadily in mind that each organic being is striving to increase at a geometrical ratio; that each at some period of its life, during some season of the year, during each generation or at intervals, has to struggle for life, and to suffer great destruction. When we reflect on this struggle, we may console ourselves with the full belief, that the war of nature is not incessant, that no fear is felt, that death is generally prompt, and that the vigorous, the healthy, and the happy survive and multiply. (129)

While he does not mitigate the truth of the matter, he puts a better face on it than Maupassant does, for the most part. Cases of the negative side of the struggle for existence far outnumber optimistic ones in Maupassant who tends to dwell on the destructive quality of both the natural and socioeconomic environments.

Natural selection

The phenomenon of natural selection is understood as a direct consequence of the struggle for existence and the adversity of the environment. It is sometimes difficult to separate the topics for discussion or to differentiate between cases of one and the other among these issues. Success or failure in the struggle for survival is in itself a matter of natural selection, and the causes that bring it about are necessarily a matter of the environment. Given all this, we shall divide the matter of natural selection in Maupassant's fiction, for clarity's sake, into topics of survival, genetic characteristic, selection against a trait, adaptation, and chance.

To what has gone before on the matter of survival in the struggle for existence, we can only add at this point that it is a question of differential advantage, as Darwin sees it. Whether in the natural environment or within the frame of human social structure, survival involves exploiting some trait—which other individuals may not enjoy—that gives an advantage against the adversity of the milieu (sparsity of resources, predators, harshness of climate, etc.); the same conditions are understood to obtain in the wild as in society. There is a very clear statement of the situation in *Bel-Ami* when Norbert de Varenne, speaking of the vicissitudes of surviving as military colonists in a foreign land and culture:

> "Oui... ils sauront tout, excepté l'agriculture. Ils parleront l'arabe, mais ils ignoreront comment on repique des betteraves et comment on sème

> du blé. Ils seront même forts en escrime, mais très faibles sur les engrais. Il faudrait au contraire ouvrir largement ce pays neuf à tout le monde. Les hommes intelligents s'y feront une place, les autres succomberont. C'est la loi sociale." (R, 216)

This is a concise statement of (Social) Darwinist principle, namely that those individuals with a helpful advantage—here, intelligence—will have what it takes to survive while others will fail. De Varenne calls it "the social law" but we recognize it as the principle of differential advantage in natural selection.

The quest for differential advantage is nowhere better illustrated in Maupassant than in M. Daron, the 86 year-old central character of "Un vieux" (CN1, 566–569). Daron comes to take the waters at Rondelis and engages a physician to report to him in great detail about all other octogenarian residents of the area, their health, the regimens they follow, and so on. The narrator reports Daron's thoughts on health and survival as a kind of applied study of advantage in the effort to stave off death. Something of an egotist, he envisions himself the *Übermensch* where natural selection is concerned:

> Il n'avait jamais considéré les autres hommes que comme des espèces de pantins créés pour meubler la nature. Il les distinguait en deux classes: ceux qu'il saluait parce qu'un hasard l'avait mis en rapport avec eux, et ceux qu'il ne saluait pas. Ces deux catégories d'individus lui demeuraient d'ailleurs également indifférentes. (CN1, 567)

Genetic characteristic is at the heart of Darwin's theory of evolution and, although he did not have Mendelian genetics in his arsenal to explain the mechanics of inheritance, he knew that evolution and natural selection hinged on the passage of advantaged traits from one generation to the next. While Maupassant does not make the matter of genetic inheritance itself a significant motif in his fiction (as he might have, for instance, in the cases of M. Daron or even "Bel Ami" himself), he does give attention to the recognition of genetic characteristic, and this betokens a level of importance for the Darwinian perspective on his literature.

A case of such awareness on Maupassant's part is his fictional rendition of the human quality of femininity. The most casual reader of Maupassant is aware that the author's usual (though not invariable) tone is a misogynistic one. So one is not surprised to see that the novels and short stories incorporate the belief that human female sexuality is dangerous in some way. Numerous narratives attest to this view. "Vains conseils" (CN1, 1213–1216) is one such text which puts forth a male epistolary narrator purporting to educate a younger friend on matters of human sexuality. His opinion is that femininity is a genetic trait for which the male of the species ought to evolve a defense, and he proposes a curricular approach:

> Mais je voudrais qu'il y eût dans les collèges un enseignement spécial pour prémunir les jeunes élèves contre les dangers de cette nature. (CN1, 1216)

Another noteworthy example of this mindset is the *style indirect libre* rendition of the thought process of Pierre in *Pierre et Jean*, once he has concluded that his mother had been unfaithful to his father with Maréchal, the father of Jean. Pierre turns stridently misogynistic, envisioning women as genetically duplicitous and manipulative as a matter of nature, which is to say, of heredity (R, 770 and 775–776). In the latter passage, Pierre is reflecting bitterly on the women at the beach and their age-old tendency to appeal to men other than their husbands; there is a very similar scene in *Notre coeur* (R, 1136) where Mariolle broods over "Leurs mères, toute celles des générations passées" and the "attrait que la forme féminine exerce sur le coeur des mâles."

Of course, natural selection effects a selection against traits which are not advantageous to survival. It is this "selection against" which, when matching characteristics of heredity or behavior against the rigors of the environment, constitutes the disadvantage of the less fit individuals. Two cases bordering on the biological may be found in the corpus. In "Une veuve" (CN1, 533–538) the trait of lovesickness is passed down genetically through generations of males in a family, and is manifestly selected against as it leads to suicide in the three cases mentioned in the narrative. The genetic tendency to alcoholism is also broached in "Le Baptême" (CN2, 436–441) which alludes intertextually to Zola's *L'Assomoir* by name (CN2, 436), and as well in "Le Petit Fût" (CN2, 77–82); in all three of these texts the genetic predisposition leads to the deaths of all involved, and hence is strongly selected against; in "Le Baptême" especially, it is a familial and tribal matter among Bretons, and hence has a dimension of biological heredity to it.

In "La Roche aux guillemots" (CN1, 411–415) we see a quite Darwinian presentation of the motif of selection against a hereditary trait. The trait in question is a natural one, the migratory pattern of a species of bird. In the text, humans change the environmental pressure on the species by hunting relentlessly at the site of the birds' migratory destination. The birds, for their part, are unable to adapt a new pattern within the time frame of the narrative, and are left in peril of extinction.

Certain aspects of behavior are aberrant, and find themselves selected against in the human social system. Here we approach Social Darwinism and its selection process. Generations of males described in the family featured in "Le Loup" (CN1, 625–630) have an inherited proclivity for hunting until it costs the life of one among them; thus selected against, the trait is resisted by subsequent generations who adapt their behavior and avoid hunting. A more comical matter arises in "Ce Cochon de Morin" (CN1, 641–652). The title character breaks from socially acceptable behavior and impulsively kisses a beautiful woman traveling in his train compartment. The woman presses charges until a friend of Morin persuades her to drop them, seducing her in the process; Morin, for his overt and unacceptable behavior, is forever known as "ce cochon de Morin," while his discreet friend experiences no selection against his conduct which is more in keeping with social custom.

The context of competition in the social setting is implicit in almost all of the corpus. It is made quite explicit, however, in "L'Héritage," specifically in the text's opening passage (CN2, 3–8) where office colleagues discuss the competition for survival within their bureaucracy. The same tone is struck in "Opinion publique" (CN1, 219–224) where office colleagues discuss office politics and goings-on in society for the duration of the text. Obviously, the social milieu in its mores and accepted values selects against poverty and those in its clutches: the perfect example is the generous central character of "La Rempailleuse" (CN1, 546–552) who, throughout her life of poverty, continually gives her savings to the bourgeois pharmacist's son whom she loves; she remains an outcast in life and in death, for her legacy of 2,300 francs to her beloved is received with outrage by him and his wife (but received nonetheless). Society is heartless in the way it leaves the poor by the wayside, but it is a fact of life which neither characters nor narrators can change.

For Maupassant's closest approximation of a theoretical basis for his Social Darwinism, we find a fairly direct reference to Spencer: in "Les Bécasses" there is the allusion to the scientific writings of a foreigner, "un brave savant de génie" who has devoted his career to the study of facts "relatifs à l'influence du fonctionnement de nos organes sur notre intelligence" (CN2, 563).

In light of the competitive side of natural selection and the unforgiving nature of the environment (either that of the wild or that of society), it is logical to ask, how does the individual respond for his or her own benefit? Adversity in the environment is overcome by adaptation, the mechanism in the behavior or morphology of beings which effects evolution in Darwin's theory. Adaptation aids the individual in the struggle for survival, in however small a way, and is thus a part of the process of natural selection, We now briefly look at the two types of adaptation in Maupassant's fiction.

The more common response in Maupassant is the adaptation of behavior to respond to the rigors of the struggle to survive. One example of behavioral adaptation is the manner of the character Clovis in *Mont-Oriol*. He is first presented thus:

> C'était un vieux paralytique, célèbre dans tout le pays où il rôdait depuis dix ans d'une façon pénible et lente... maintenant il geignait et déambulait à la manière d'un crabe qui auarait perdu ses pattes. Il allait, traînant par terre la jambe droite comme une loque, et la gauche relevée, pliée en deux. Mais les garçons du pays...affirmaient qu'on rencontrait le père Clovis, rapide comme un cerf et souple comme une couleuvre sous les buissons et dans les clarières.... (R, 523)

Clovis has evolved a manner of behavior which enables him to beg for a living and pass for indigent when society is looking; for the rest, he continues his life as a poacher. His affected limp and disability have the later benefit of enabling him to pose (contractually and for money) as one cured by the waters of the spa which investors make of Mont-Oriol.

We have already considered the conduct of the title character in "L'Aventure de Walter Schnaffs" (CN1, 792–801). He comes up with a strategy for surviving the shot and shell of battle. His motivation is largely that of self-preservation, but it has a dimension of the concern for progeny that often accompanies conduct in Darwinian study: Schnaffs worries about his children's fate should he be killed (CN1, 793). The plan he hits upon is to surrender at any given chance.

"La Confession de Théodule Sabot" (CN1, 1021–1029) illustrates another behavioral adaptation for advantage. Théodule the carpenter spends his life as an anti-clerical miscreant, but when a lucrative contract becomes available at the town church, Théodule confesses and converts, and true to his word, he remains a practicing churchgoer.

Another subject of some levity is the adaptation required of Blérot in "Un sage" (CN1, 1087–1093). This friend of the text's narrator has a peculiar problem: his wife's romantic ardor is bringing him to a point of physical exhaustion and is beginning to compromise his health seriously. The narrator, to save his friend's life, counsels him to curtail his sexual activity and to arrange for the wife to take lovers. Blérot follows the advice, and in the end everyone is happier.

Again we may consider "Le Vagabond" (CN2, 856–867) whose title character is forced to adapt to life on his own in the outdoors, in order to stay alive. He takes shelter and nourishment from a cow in the field, he finds shelter and food elsewhere as he can, and he is forced to a life of petty crime in the face of society's inability to provide work or sustenance. This adaptation, however, is repressed by society.

"La Parure" (CN1, 1198–1206), falls under this rubric. Mathilde Loisel is born of the petite bourgeoisie but ever suffers under the sentiment that with her natural beauty she deserves more of life; when her husband procures an invitation to a high society ball, Mathilde borrows a necklace from a wealthy friend; upon losing the necklace after the ball, Mathilde and her husband spend ten years of drudgery working to pay for the replacement, only to find in the end that the original necklace was fake. Mathilde is presented in the opening sentence as one of a sub-species ("...une de ces jolies et charmantes filles, nées comme par une erreur du destin, dans une famille d'employés" CN1, 1198) that is poorly adapted to its environment—in this case, unwilling to accept the circumstances of life in which she is born. Darwinian theory leaves one choice: either adapt or face extinction at the hands of a blind (i.e. society does not sufficiently see her beauty, etc.) and pitiless environment. Mathilde effects camouflage to enhance her standing (by purchasing a *toilette* for the ball with money apportioned for something else, and by borrowing the necklace), and herself falls prey to part of it, since the necklace turns out to be camouflage (costume jewelry). Driven by moral sense, out of tune with the amorality of the environment, Mathilde Loisel (*l'oiseau*) migrates to the periphery of her range (the couple moves house, "...on loua sous les toits une mansarde" CN1, 1204), and like one of Darwin's Galápagos finches, she is forced by harsh conditions to adapt and pay the bodily price of survival. Ten years of adapting to the struggle to survive leave Mathilde haggard in body and spirit, bereft of the beauty she had considered her entitlement.

The other variety of evolutionary adaptation is anatomical, and there is only one case in the corpus which approximates it. In "La Mère aux monstres"(CN1, 842–847) the title character attempts successfully to effect anatomical aberration in her offspring in order to gain financially. This is the horrifying tale of a woman who inflicts injury upon herself during pregnancy, so that she might give birth to freaks whom she would sell for exhibition. The narrator is reminded of his encounter with "cette horrible femme" when he sees children on the beach whose deformities are the result of their mothers' having worn corsets during pregnancy for beauty's sake. These are incidents of physical change, brought about not for evolutionary advantage, but as either a means of financial gain or as the unintended consequence of adaptive behavior. For these reasons, they do not qualify precisely as anatomical "adaptations" in themselves, but they are as close as Maupassant comes to including the motif in his fiction.

Beyond the phenomenon of adaptation, there remains one topic for consideration under the rubric of natural selection, the question of the role played by chance in the survival of individuals and species. In a strictly Darwinian reading, chance applies more or less exclusively at the microbiological level—the random mutation of genetic information in the reproduction process. We do not speak here of this kind of random chance, as I do not know of any reference to microbiology in Maupassant other than the passing reference in *Notre coeur* to "...la découverte la plus récente, incertaine encore, des microbiologistes" (R, 1039). As a general topic, the mutation itself of genetic characteristic does not seem to enter into the novels and short stories.

There is, however, a motif of random chance aiding the individual in the general struggle for existence. One clear example of this is the central motif of action in "Une page d'histoire inédite" (CN1, 186–192), the story of random events which preserved the life of the young Napoleon Bonaparte. As the narrator phrases it, it is "ce grand hasard des événements qui gouverne les hommes et le monde, un fait bien petit..." (CN1, 186), exactly the sort of event which, in Darwinian perspective, has a great effect on survival and thus on the world. In this case, the gesture of a woman which detains her husband a moment permits Napoleon to escape, and changes the course of European history.

Other such occasions, though less vivid, in the corpus include the chance meeting of Georges Duroy with his old acquaintance, Forestier, at the beginning of *Bel-Ami*; this meeting affords Duroy the chance to rise out of poverty and launch the career that changes his entire life. There is also the realization on the part of *l'académicien* in "Un fils" (CN1, 416–425) that his own neglect denied his bastard son the chance to grow up in different surroundings, which certainly would have changed the son's life entirely for the better.

The "arms race"

One of the most fascinating phenomena observed by evolutionary science is known as the arms race. This is the escalating evolution of anatomical or behavioral weapons by competing species or by competing members of the same species. It is analogized to an arms race because it is understood to involve a move-and-countermove relationship over evolutionary time, necessitating an evolutionary response to each advantageous development in the opposition (be it in another species or among members of the same species competing, say, for preference in reproduction). The race is often costly to both sides and usually ends in a standoff or in a truce of coexistence. The arms race phenomenon is thought, for example, to be the best explanation for a forest's canopy being of uniform height: all trees compete for sunlight, yet unending growth would be too costly for the organisms to sustain; if trees strike a balance in height, then all can have access to sunlight without the expense of further growth. The arms race is also behind the evolution of greater toxicity in prey to offset evolving immunity in predators.

For a thematic illustration of the "arms race," we need not look very far in Maupassant's fiction. There are humorous examples of the escalating standoff, such as in "Un coup d'état" (CN1, 1004–1014), where the passage from Empire to République is marked in the tiny town of Canneville by a siege of the town hall. And there is the competition in "Le Remplaçant" (CN1, 700–703) between two soldiers to be the paid consort of the widow Bonderoi, which ends in a truce of assigned days.

More serious arms races are detailed in the corpus, too. The competition for economic advantage among capitalists and investors in *Mont-Oriol* is mirrored in the redoubling of efforts on the part of competing physicians to attract clients. But where the rivalry seems to take on truly evolutionary aspects is in the battle between the sexes portrayed throughout the novels and short stories. As prospective lovers endeavor to seduce and to overcome the resistance of their intended partners, as rivals try to outdo one another for the attention of the one whom they would have, stakes increase and the struggle can be costly (in money or in efforts). The text of "Vains conseils" (CN1, 1213–1216) is a letter from an older advisor to a young man alerting him to the evolved arsenal of feminine weapons in the arena of love, and expressing a wish for a masculine countermove of increased awareness and defense. The link to the evolutionary model is underscored here by the fact that the conflict in question has to do with the reproductive function and also (ostensibly) involves a response on the part of the males of an entire species to evolved traits in the females of their species. The letter writer is aware of the expense of the "arms race" he proposes, but has weighed those expenses against the risks he perceives.

Social and economic ends in reproduction

This subject is included in the present chapter only by virtue of its association with Social Darwinism, an outgrowth of the latter understood in its broadest, and at times, its most radical applications. When reproduction is viewed as a vehicle of social engineering, the application of principles ascribed to Darwinism has become known as "eugenics," defined as the proposed improvement of heredity by the control of parentage chosen for its presumed genetic desirability according to standards of skin color, ethnicity, intelligence, and so on. Founded as a school of thought by Francis Galton (1822–1911), a cousin of Charles Darwin, it went through various phases of acceptance in Europe (culminating in the atrocities of the Nazi regime) and the United States (where it produced anti-miscegenation laws and forced sterilizations in many states through the mid-twentieth century). The movement, of course, has nothing to do with Darwin; it has only to do with some of the ingredients of Darwinian theory—reproduction, heredity and survival—but not with evolution or natural selection. Still, it has come to be linked to Social Darwinism for the latter movement's harshness and inclination to social engineering, and because Darwin used in his *Origin* the illustration of choosing desired traits in the planned breeding of domesticated animals as a metaphor for heredity in natural selection.

Maupassant does not show any favor for eugenicist thought in his fictional thematics. The inverse of the topic comes up in "Berthe" (CN2, 355–364), the narrative telling (ostensibly in the voice of her physician) of a young woman, severely underdeveloped mentally, and how her caretakers decide that marriage would be the best therapy for channeling her passions. The experiment ends in disaster when her husband becomes abusive. This is marginally related to eugenics in that it involves the sexuality of the mentally underdeveloped and decisions made clinically by others to control them; sterilization, however, is not in question (in fact, Berthe's father envisions maternity as stirring her intelligence, hitherto inert, when he proposes the idea of Berthe's marriage to the physician/narrator, CN2, 360).

There are two cases of parentage arranged for financial gain in the short stories. Both involve substituting a fertile male partner for a sterile husband, in order to satisfy the demands of a will and to effect the inheritance of a fortune. One is in "L'Héritage" (CN2, 3–69) and the other in "Un million" (CN1, 614–619). In both texts, the ending makes the hypocrisy of a bourgeois, moralistic wife a matter of implicit ridicule, more or less diverting attention from the question of engineered extramarital parentage.

These, with the exception of the appendix on heredity engineering, are the central topics of evolutionary theory. The hostile environment, the amorality of nature and the related cruelty or altruism in the circumstances of life, the struggle for existence and the related principle of natural selection, and the Social Darwinism which applies evolutionary theory to human society, all combine to form the core of Darwinian theory. Their inclusion in the thematics of Maupassant's fiction to the extent we have observed attests to the author's global awareness of evolutionary theory, as well as to his understanding of its application to nature and to society.

From here we move on to the study of the very vehicle of adaptation and species change, the ensemble of subjects centered on sexuality: sexual selection, parentage, reproduction and heredity.

NOTES

1. Two notable exceptions to this , both of which decry a God who is the creator of death and disease, are "L'Inutile beauté" discussed in Chapter 4, and the end of the fragment *L'Angélus* (R, 1223).
2. Paul Lefebvre, "'Le Horla': Une oeuvre de Guy de Maupassant (1850–1893)," www.denise-pelletier.qc.ca/fiches/pieces/horla.html.
3. They are "Sur l'eau" (CN1, 54–59), "Conte de Noël" (CN1, 689–694), "Lui?" (CN1, 869–875), *Pierre et Jean* with its references to "l'autre qui est en nous" (e.g. R, 736), "La Chevelure" (CN2, 107–113), "La Peur" (CN2, 198–205), "Un fou?" (CN2, 308–313), "Lettre d'un fou" (CN2, 461–466), "Un fou" (CN2, 540–547), "Le Horla" first version (CN2, 822–830), "Le Horla" second version (CN2, 913–938), and "Divorce" (CN2, 1017–1025).
4. See Ropars-Wuilleumier, ed., preface to Guy de Maupassant, *Pierre et Jean* (Paris: Albin Michel, 1984) 8–9.
5. See CN1, 1498, note 2.
6. See Stephen Jay Gould, "Nonmoral Nature" in *Hen's Teeth and Horse's Toes* (New York: Norton, 1983) 32–45 for an excellent, lucid essay on the topic of amorality.
7. See the research of Sharon P. Johnson, *Boundaries of Acceptability: Flaubert, Maupassant, Cézanne and Cassatt* (New York: Peter Lang, 2000) 135–163, for a fine study on the topic of rape and bourgeois mores in "Madame Baptiste."
8. See the works of Richard Dawkins, *The Selfish Gene* (New York: Oxford University Press, 1989) and *River Out of Eden: A Darwinian View of Life* (New York: Basic Books, 1995) whose thesis is that the genetic imperative of self-propagation provides an impetus for altruism towards kin. This topic will get further attention in chapter six.
9. E.g. "Yveline Samoris" (CN1, 684–688).

Chapter 6: Reproduction and Sexual Selection

Certainly sex and sexuality are recurring thematic *topoi* in Maupassant. Writing about sex does not make one a Darwinian, and much of the erotic interest in Maupassant's fiction will not be pertinent to the present study. What will figure in our reading will be the thematic aspects of sexual relationships and sexuality that have to do with sexual selection and the passing along of a genetic heritage.

It is assumed today that Darwin recognized the putative criticism of natural selection as a lone force shaping evolution; perhaps as a way of deflecting some of that criticism he proposed sexual selection as another force driving evolutionary development. In any case, many evolutionary scientists now recognize sexual selection as one kind of natural selection. Sexual selection involves the dynamics of sexual attraction and choice of partner, the courtship advantages of one individual over the other (efficacy of sexual lures, perceived desirability, etc.) as well as the competitive advantage of one individual over the other in a rivalry over a prospective partner. In evolutionary theory, sexual selection pertains mainly—though not entirely—to the males of species. As Darwin phrased it in the chapter subsection in the *Origin* dealing with sexual selection,

> This depends, not on a struggle for existence, but on a struggle between the males for possession of the females; the result is not death to the unsuccessful competitor, but few or no offspring. (136)

Despite Darwin's vocabulary and circumscription, evolutionary science has always paid attention to the active female role in sexual selection.

Human sexuality and that of Maupassant's characters is far more complicated than even the Darwinian take on the subject, as a function of human culture. Fortunately there is the insightful research of

Sharon P. Johnson which goes well beyond the usual dimension of Darwinian perspective on sexuality to situate the motif in its cultural and psychological scope. This study, rooted in Foucault's theory of power and sexuality, details the subversion of traditional representations of gender and class where sexuality is concerned (1–8 and 135–163).

Selection aside, the mark of success in the view of evolutionary theory is reproduction, the passing along of one's genes to a new generation. The point of evolving adaptations for success in the struggle for existence is not so much survival in itself (though that may be rewarding to the individual), but instead procreation (of genetic interest to the species) which assures the replacement of the individual on his or her eventual death, as well as the continuance of the genetic line. In this chapter we will include all topics related to reproduction and sexuality in an evolutionary perspective: the biology of reproduction, impulse and inhibition in sexual conduct, incest, sexual lures and rivalry, heredity and parentage, reproduction as evolutionary success, and adoption.

The biology of reproduction

Sex and reproduction are, of course, nothing new when it comes to literary motif. However, the Naturalists bring a new interest in life processes and biology that focuses narrative attention on the body to an extent not often seen in literature up to the nineteenth century. Maupassant shares the interest in this vision, and his fiction does not shy away from realism where the biology of reproduction is concerned. This includes the motif of abortion which is set forth openly (e.g. "L'Enfant" CN1, 981–986, which begins with the phrase "On parlait, après le dîner, d'un avortement qui venait d'avoir lieu…). The central action of "En wagon" (CN2, 478–484) involves a group of schoolboys and their chaperone priest who aids in the delivery of a baby in a train compartment. In *Une vie,* the birth of Rosalie's baby is described in some detail (R, 79).

One character is led by his thoughts to speak of the dynamics of reproduction biology, and to draw a parallel between trees and humans:

> "Quand on songe que ces imperceptibles atomes qui sentent bon, vont créer des existences à des centaines de lieues d'ici, vont faire tressaillir les fibres et les sèves d'arbres femelles et produire des êtres à racines,

> naissant d'un germe, comme nous, mortels comme nous, et qui seront remplacés par d'autres êtres de même essence, comme nous toujours!" (CN1, 416)

The metaphorical juxtaposition of human and plant sexuality recurs in Maupassant, for example in *Une vie* where the season of spring "...semblait remuer les sèves chez les hommes comme chez les plantes" (R, 116). As for the marvelous regeneration of life, the central character of "Le Baptême" (CN1, 1144–1148) is moved by the presence of a newborn to reflect on the "mystère ineffable" of new life,

> ...le grand mystère de la vie qui commence, de l'amour qui s'éveille, de la race qui se continue, de l'humanité qui marche toujours. (CN1, 1147)

Such a motif calls attention to life as an object of biological interest, as well as to its property of continuity which appears to lie in the genes rather than in the faculty of reason.

Of course, reason and all that comes with it (culture, morality, etc.) occasionally comes into conflict with the biological imperative to reproduce, leading the reader and even the narrator to wonder if the genders' different priorities are acquired or written in the genes. The epistolary debate in "Les Caresses" (CN1, 952–956) between male and female correspondents speaks for itself as to the sexes' divergent viewpoints on the appropriate frequency of sexual contact. One facet of this motif is also evoked in *Une vie* at several points (e.g. where the story of the baker's wife romantic tryst in the large oven puts Jeanne off bread for a while, R, 116–117). In any event, if a generality is to be drawn about Maupassant's fictional representation of the biology of reproduction, it is that this aspect of life is treated as an integral part of that life, and one that is not to be hidden for the sake of *bienséance*. It is part of the organic continuum of life, and attention to it gives evidence of interest in a naturalist's (with small n) view of nature and the world.

Natural impulse vs. morality

Since reproduction is the necessary mark of success in evolution, given that survival of the individual serves evolutionary ends only insofar as the individual is successful in leaving progeny, the imperative of life is reproduction. While the sex drive was the target of repression in Classical aesthetics and was idealized in Romanticism, to the Naturalists it was more simply a fact of life. The influence of

Darwinian thought is to make vivid the contrast between the amorality of nature, the postulate of Darwinism that nature pursues its own ends without regard for other protocols, and the dictates of a moral order of cultural origin. On the one hand, the needs of life are to reproduce and to assure continuity of the genetic line. On the other hand, morality calls for keeping the passions in check. The disparity brings impulse into conflict with concern for propriety, where impulse is nature's response to the need for propagation and restraint is morality's commandment.

The motif of sexual pursuit and interplay recurs almost countless times in Maupassant. Sometimes it is lighthearted, sometimes it causes sorrow or disaster. Rarely is it portrayed as evil (as it is with rape or brutishness), and always is it a part of the human condition, a function of human nature to the extent that humanity represents an animal species like any other. This section of the study is not an inventory of the motif of sexuality in the corpus, but instead a detail of the instances of the motif which relate to a Darwinian reading. For this reason we focus on the depiction of sexuality: for the most part, that depiction is in keeping with the overall Darwinian perspective that humanity constitutes a species of life in a natural, earthly environment that itself is unaffected by moral scruple; on occasion, this view is set in contrast with a motif of moral concern in characters.

The contrast is most evident in *Une vie* where the central character and her beloved father have diverging opinions about sexuality; they never debate the issue directly, but the contrast in their opinions is evident. Jeanne comes to a distaste for the matter partly out of revulsion for her husband's excesses and lack of discretion, and partly out of what appears to be naïve idealism about sexuality in love. As early as her honeymoon, Jeanne feels disaffected:

> Elle ne disait plus rien, les yeux baissés, révoltée toujours dans son âme et dans sa chair devant ce désir incessant de l'époux, n'obéissant qu'avec dégoût, résignée, mais humiliée, voyant là quelque chose de bestial, de dégradant, une saleté enfin. (R, 53)

It is in fact the animal side of sexuality that offends Jeanne's sensibilities. Jeanne's personal inclination is to separate the physical from the spiritual in the way that Platonic influence led aesthetics to do up through the literature of the pastoral tradition. The very aspect of sexuality that links it to evolutionary thought is what Jeanne finds offensive, though note that it is not nearly so much an actually moral judgment as an aesthetic one. Later in the novel when Jeanne falls for a while under the sway of the scrupulously anti-sexual Abbé Tolbiac, it is for sentimental and not spiritual reasons, for as the narrator says, "La religion de Jeanne était tout de sentiment" (R, 135). It is something of an anomaly for Jeanne, since later in life her only consolation in the world is her son Paul, the fruit of her sexual relationship with a brutish husband. Sexuality for her remains a matter of "...répugnance qui devenait haineuse, de cette sale bestialité." Her attitude is elaborated:

> L'accouplement des êtres l'indignait à présent comme une chose contre nature; et, si elle en voulait à Gilberte, ce n'était point de lui avoir pris son mari, mais du fait même d'être tombée dans cette fange universelle.
>
> Elle n'était point, celle-là, de la race des rustres chez qui les bas instincts dominent. Comment avait-elle pu s'abandonner de la même façon que ces brutes? (R, 116)

Her disposition, strong though it is, is born of distaste rather than moral objection. In this she is a little like Christiane of *Mont-Oriol* whose renunciation of ideals of love is based not on moral grounds of objection, but on a belief that she and her beloved

> ...s'étaient seulement un peu rapprochés jusqu'à faire toucher les impénétrables enveloppes où la mystérieuse nature a isolé et enfermé les humains. Elle vit bien que nul jamais n'a pu ou ne pourra briser cette invisible barrière qui met les êtres dans la vie aussi loin l'un de l'autre que les étoiles du ciel. (R, 691–692)

It is "mysterious nature," and not moral injunction, that gives rise to Christiane's disinclination.

For objection to sexual activity on moral grounds, one would have to turn to either bourgeois hypocrisy or clerical adherence to moral law. We have already alluded to examples of both: the wives in "Un million" and "L'Héritage" who, once having gained wealth by giving birth despite their husbands' sterility, become judgmental of other women, and the two priests (l'abbé Tolbiac in *Une vie* and the zealous priest in "Le Saut du berger") who, in identical episodes, slaughter a dog and its pups out of rage over the fact that children are watching the birth process. Two other priest characters, in similar circumstances, sternly disown their nephews when catching them in sexual dalliances ("Une triste histoire," CN1, 153–158 and "Une surprise" CN1, 816–822). There is a pattern here: moral objection to sexual activity is routinely pictured in Maupassant as extreme and unreasonable, itself out of step with nature.

Instead, the prevailing narrative attitude towards sexuality is that it is a function of nature and, in itself, unobjectionable. In that it serves the need of the species to assure its future, it is recognized as a life process, a biological reality of existence, and as such is not particularly subject to questions of morality. Legal questions occasionally arise, as in "Au bois" (CN2, 758–762) when a 60 year-old Parisian bourgeois couple is arrested for indulging the woman's romantic nostalgia and engaging in sex in a forest; the mayor lets them off with a warning. Again, the issue is not one of good or evil, but legal or ethical.

To return to *Une vie,* it is the naturalist father of Jeanne who gives voice to the attitude towards sex and reproduction that is more typical of the corpus. Of the baron, the narrator says,

> Il était, lui, de la race des vieux philosophes adorateurs de la nature, attendri dès qu'il voyait deux animaux s'unir, à genoux devant une espèce de Dieu panthéiste et hérissé devant la conception catholique d'un Dieu à intentions bourgeoises.... (R, 138)

His view of nature is of an infinity

> ...plus vaste qu'un raisonnement, produisant sans but, sans raison et sans fin dans tous les sens et dans toutes les formes à travers l'espace infini, suivant les nécessitees du hasard.... (R, 138)

One would be hard pressed to find a more Darwinian expression describing nature and its procreative force.

Human sexuality, then, is a reflection of nature's protocol. Love is a consideration apart, having nothing inherently to do with the process of physical procreation. According to Bertin of *Fort comme la mort,* one thinks of a beloved as

> ...une femme ou un homme, une créature sans nom, sortie de la Nature, cette grande femelle, avec des organes, une forme, un coeur, un esprit, une manière d'être générale qui attire comme un aimant nos organes, nos yeux, nos lèvres, notre coeur, notre pensée, tous nos appétits sensuels et intelligents. On aime un type.... (R, 966)

giving evidence of the materialistic view of life (and even love) that is at the heart of Darwinian theory.

The journalist of "Un cas de divorce" (CN2, 777–783) gives his impression of the workings of nature:

> Deux bêtes, deux chiens, deux loups, deux renards, rôdent par les bois et se rencontrent. L'un est mâle, l'autre femelle. Ils s'accouplent. Ils s'accouplent par un instinct bestial qui les force à continuer la race, leur race, celle dont ils ont la forme, le poil, la taille, les mouvements et les habitudes.
>
> Toutes les bêtes en font autant, sans savoir pourquoi!
>
> Nous aussi. ... (CN2, 780)

Again, a more Darwinian description of life processes and human ontology one could not find. Humanity is explicitly placed on a par with other animal species, and all species are seen to respond to the same imperative to reproduce in an amoral context. Responses to the sex drive are essentially the same across the horizon of animal species; the narrator of *Fort comme la mort* links Bertin's desire to be attractive to young Annette to the "envies instinctives...qui font faire la roue aux paons et des vers aux poètes" (R, 975).

When the baron in *Une vie* explains the facts of life (or the "choses de l'existence") to his daughter Jeanne, he describes the sexual bond between spouses as "la loi naturelle" (R, 44). Later, the narrator records the baron's thoughts setting "les lois et les instincts innés" in opposition to the moral "persécution" carried out by the abbé Tolbiac (R, 138). The order of nature is couched in the vocabulary of law, and the opposition of human law to natural law is not gratuitous. The latter clearly takes precedence over the former throughout the corpus as a matter of implicit textual thesis. Bereft of romantic trappings, sexuality involves "l'éternelle lutte entre les deux brutes humaines, le mâle et la femelle" (CN2, 1102), and in its innocent form, as in its most guilty manifestations, it is "un emportement bestial" (CN2, 639), "une sorte d'amour bestial" (CN2, 641).

If there is resolution to the debate between nature and morality in the fiction of Maupassant, it is in the short story "L'Enfant" (CN1, 981–986) where the physician/secondary narrator pleads the case of a young woman whose strong sex drive is described as a natural and hereditary trait; the physician chides bourgeois morality for holding the young woman guilty. Hereditary chance has left this trait in some, and "...les sens sont invincibles" (CN1, 982). The physician asks,

> Pouvez-vous arrêter le vent, pouvez-vous arrêter la mer démontée? Pouvez-vous entraver les forces de la nature? Non. Les sens aussi sont des forces de la nature, invincibles comme la mer et le vent. (CN1, 982)

It may be the decree of "la perfide nature" (CN1, 982), but it is the order of nature nonetheless, so those who speak for morality should not sit in judgment.

There is, however, one form of sexual expression which meets with censure in Maupassant and in morality alike. It is the phenomenon of incest, and it is worth noting that biology as well has long joined in the chorus of reprehension, if only for reasons borne out in the study of genetics since Darwin (but suspected in principle well before that). Particulate genetics has shown that the likelihood of genetic defect is higher when similar genes are combined in reproduction; conversely, that probability decreases when divergent genes are combined (as the defects in one are less likely to be matched in the combining gene). Biology advises against incest, and Maupassant appears to do the same.

There are three short stories in which incest is the central motif. Two of these involve relations between father and daughter. In "L'Ermite" (CN2, 685–691) the narrator encounters a man who has retreated from active social life after discovering his accidental incest: after a casual encounter with a younger woman, he spots on her mantle a daguerreotype of himself as a young man; he inquires about the image and is told by the unsuspecting woman that her mother had left it to her as the picture of her father. In "Monsieur Jocaste" (CN1, 717–721) the protagonist Pierre Martel falls in love with his long lost daughter, seeing in her the image of her mother whom he had loved; the young woman, for her part, falls in love with him. The narrator of this tale, however, mitigates the guilt of "M. Jocaste," saying "Que peut...la pudeur contre l'irrésistible volonté de la nature?" (CN1, 718) and ascribing the younger woman's birth in the first place to "Le hasard des fécondations, la loi brutale de la reproduction" (CN1, 720–721). The narrator recognizes that Jocaste is not 100 percent certain he is the woman's father, and that Martel acts out of a sense of duty (i.e., his promise to the mother to give his life to the child, even at the price of a crime) when he marries the young woman. The question of offspring is skirted in the narrative, so the genetic aspect of the problem is disposed of. The narrator's interest is, as he states to his interlocutor at the outset, not to excuse "le fait immonde et brutal" of incest, but to show that "...on ne peut pas lutter contre certaines fatalités qui semblent des fantaisies horribles de la nature toute puissante." (CN1, 717) Incest here is, like everything else in life, the handiwork of nature.

In "Le Port" (CN2, 1125–1133) a seafaring man discovers too late that he is with his long lost sister in a brothel. His reaction is one of horror and leaves no doubt as to the implicit narratorial stance on the matter of incest. In all of the cases cited, incest is the product of chance and fate, the fate of nature which remains amoral. Incest is nonetheless treated as a taboo, a regrettable accident of nature.

Sexual selection, lures and rivalry

It was mentioned above that Darwin viewed sexual selection as a phenomenon first involving rivalry among males within a given species. His observations led him to conclude that competition for mating goes on primarily between males, and that females tend to select according to the outcome of such competition. This process, in somewhat nuanced fashion, works out on occasion in Maupassant; at other times women take a more active part in the competition involved in selection than Darwin notes in species in general.

We have already discussed some examples of selection in itself, where in two short stories, a woman chooses one of her husband's co-workers to impregnate her (since the husband is evidently sterile). In each text ("L'Héritage" and "Un million") dialogue between the spouses makes an issue of the husband's unsuitability for procreation (that is, of the wife's dissatisfaction with this), and of the preferability of the other man. In any event, the success of the reproductive effort at the genetic level has a correlative on the financial level in these texts. Thus the suitable mate is overtly chosen by the woman, and the rewards are made obvious.

In "Le Cas de Mme Luneau" (CN1, 963–967) and elsewhere, we see the motif—comically treated—of a woman who has arranged to pay for efforts of several men to get her pregnant. Again, money is substituted for the genetic reward of leaving progeny, the latter being the purely Darwinian incentive for reproduction, the former perhaps serving as a metaphor for that incentive.

Other texts involve selection by the male. "La Porte" (CN2, 902–907) shows the process by which a young man is dissuaded from his initial attraction to a woman: the woman's husband arranges for the prospective lover to be in front of the woman's dressing room door while she is dressing; when, by the husband's plan, the door is unexpectedly opened, the young man sees the woman unadorned, and decides not to pursue her further.

In "Hautot père et fils" (CN2, 1056–1068) a son takes on the care for his father's mistress when his father dies. He also inherits an attraction to the woman and selects her sexually.

Sexual lures play a great role in Darwin's view of the reproductive process, and of course they figure greatly in the fictional courtships in the corpus. For the most part, though not entirely, they are the lures of women, but there is the occasional instance of male "display" for sexual attractiveness (going beyond the obvious attention to appearances in dress, as in Bertin's haberdashery in *Fort comme la mort* or Duroy's repeated self-study in mirror-reflected images in *Bel-Ami*). The most notable example of this is in "La Moustache" (CN1, 918–922), ostensibly a letter between woman friends. The writer opines that the male moustache is absolutely indispensable for romance and sexual attraction. She says, "D'où vient donc la séduction de la moustache, me diras-tu? Le sais-je?" (CN1, 919), and proceeds to give her detailed feelings on the question, precisely in terms of attraction and courtship: the text thus poses as a subjective account of this very particular aspect of male sexual display. The importance of the male's physical attractiveness is underscored by the comment in *Fort comme la mort*, "...aucune femme, bien qu'elles prétendent, n'[est] indifférente à la beauté physique..." (R, 848).

As for the corresponding motif on the female side, there is no shortage of examples. The closely described beauty of characters in novels (Christiane in *Mont-Oriol* and Michèle de Burne in *Notre coeur*, and Annette in *Fort comme la mort*, to name only three), along with the effects of that beauty on males, is sufficient illustration of this certainly non-exclusively Darwinian motif. Maupassant tends[1] not to dwell on particular aspects of anatomy in his rendition of the process of sexual attraction, though there do appear in the short stories certain references to, for example, "la face secrète" ("La Fenêtre," CN1, 901) for the posterior, or "...de pareilles choses sous la robe de laine de la patronne" in reference to the landlady's "superbe poitrine" in "La Patronne" (CN2, 75).

As was mentioned above, Darwin's own view of sexual selection involved principally the rivalry among males competing for success in reproduction. Rivalry among men for the attentions of a woman is, of course, a recurring motif in the fiction of Maupassant. Given the title and the thematic outlay of *Pierre et Jean*, it would come as no surprise that the two principals enter into conflict over the courtship of a woman. It is the young widow, Mme Rosémilly.

> Les deux fils, à leur retour, trouvant cette jolie veuve installée dans la maison, avaient aussitôt commencé à la courtiser, moins par désir de lui plaire que par envie de se supplanter. (R, 720)

From the start, it is clear that there is more of an affinity between the woman and Jean, to the exclusion of Pierre; Jean and Mme Rosémilly are of similar temperament, share the same values in all matters of conversation that come up, and both happen to have blond hair:

> Déjà elle semblait préférer Jean, portée vers lui par une similitude de nature. Cette préférence d'ailleurs ne se montrait que par une presque insensible différence dans la voix et le regard, et en ceci encore qu'elle prenait quelquefois son avis. (R, 720)

Pierre is the odd man out, and ultimately he is the loser in courtship as in everything else. Jean ends up with the prize of Mme Rosémilly's hand and the concomitant prospects of having progeny, where Pierre is constrained to leave the "territory" and to travel off in search of better fortunes elsewhere, and he does so after giving copious evidence in the narrative of a decidedly anti-feminine sentiment. The rivalry between the brothers is evident from the opening of the novel, when the two engage in a rowing contest during a family outing in the company of Mme Rosémilly:

> ...une lutte commença pour montrer leur vigueur. Ils étaient venus à la voile, tout doucement, mais la brise était tombée et l'orgueil de mâles des deux frères s'éveilla tout à coup à la perspective de se mesurer l'un contre l'autre. (R, 723–724)

The "male pride" referred to comes into play, of course, when it comes to competing for the attention of the young widow, and both men attempt to show off in this feat of strength in order to impress her. Pierre, after a strong start, is embarrassed in the end when his strength gives out, and "Mme Rosémilly semblait ne pas voir, ne pas comprendre, ne pas entendre." (R, 724), but of course she understands perfectly what is going on in front of her: rival males are vying to impress her as suitable prospective mates.

The motif of rivalry for sex comes up again in "L'Ivrogne" (CN2, 94–99). Jérémie is detained at the bar by his "friend" Mathurin whose brother is elsewhere; the answer is proffered "Il est au chaud, t'inquiète pas" (CN2, 96) when the barkeeper asks where the friend's brother is. When a drunken Jérémie finally returns home, he detects that someone steps over him while fleeing the house (obviously Mathurin's brother). A Darwinian reading of the motif here would find that Jérémie's alcoholism has rendered him an unfit mate, and another steps into the void of sexual adequacy left by Jérémie. The text ends violently, as sexual possession is not to be trifled with: Jérémie's rage leads him to kill his wife.

According to the thoughts of Bertin, protagonist of *Fort comme la mort*,

> Tout possesseur de femme est un rival. C'est un mâle satisfait que les autres mâles envient. (R, 967)

It is thus a part of male sexuality to compete for the sexual possession that satisfies the reproductive drive, and that competition happens throughout Maupassant's fiction. In both "Les Dimanches d'un bourgeois de Paris" (CN1, 122–174) and "Souvenir" (CN2, 120–126), the protagonist supplants a feckless husband while the latter goes off in search of a lost dog and a lost wallet; in "Souvenir" the other man wins the wife's sexual favors. In both *Bel-Ami* and "Le Vengeur" (CN1, 1053–1058) a widow's second husband denigrates the dead man in an effort to bolster his own self-image and to assert his sexual superiority. In "Petit soldat " (CN2, 485–491), two best friends come into rivalry for the affection of a young woman they meet each Sunday on their walks in the country; the drive to possess her trumps their great friendship, and the loser's sense of betrayal drives him to suicide.

Sexual rivalry is not entirely limited to the male of the species. "La Confession" (CN1, 1035–1039) relates the jealousy of a girl for her older sister's fulfillment that leads her to poison the sister's fiancé. Of course, in this case (as well as in "L'Ivrogne") it is the object of sexual possession, and not the rival, that meets with death. From a Darwinian perspective, this outcome defeats the purpose of rivalry, so this development is a decidedly human turn for biological protocol.

But let us for a moment limit the discussion of sexual selection to its genuinely Darwinian bases. Evolutionarily speaking, Darwin's assumption (borne out remarkably by the science of genetics of late, as reported by the research of Spencer Wells) was that sexual selection—as a factor in the general selection process that shapes the evolution of species—is a selection exercised by females subsequent to rivalry among males: males compete to be selected, but females select. Hence feminine autonomy is (or should be) the natural order governing the life history of species. In species in which sexual reproduction is carried on by a method of competition and choice, the female selects solely by her own criteria: physical promise (for material support or support in child rearing, reproductive ability, genetic heritage for her young to profit by, etc.), beauty, intelligence, culture, and so on. The male displays his presumably desirable traits and competes for attention in hopes of being selected.

Let us leave aside for now the many (and they are many!) instances of the motif of extramarital infidelity in Maupassant's fiction, in order to consider the purely Darwinian aspects of sexual selection; the motif of infidelity in marriage is so charged with cultural significance (unrelated to the biological function of reproduction and choice of partner) that it is, in that sense, extraneous to the immediate issue. The corpus of Maupassant's narratives gives an array of examples in which Darwinian sexual selection, in its three principal manifestations, is the cardinal motif.

First, there are numerous instances of male rivalry with selection effected by women; we had occasion to discuss just above the texts of *Pierre et Jean*, "Petit soldat" (CN2, 485–491), "Un million" (CN1, 614–619) and "L'Héritage" (CN2, 3–69) in this light. There are also notably "Marroca" (CN1, 367–376), "La Martine" (CN1, 974–980) and the central action of *Notre coeur* which all display the motif. In all of these works, males undergo the rigors of competition for sexual selection and, depending on circumstances, either enjoy the rewards of being selected or suffer for losing out.

Next are the instances of male display. This evolutionary function is in evidence in *Pierre et Jean*, for example (notably in the rowing competition in the opening passage, but also in Pierre's recurring desire to outshine his brother in conversation in the presence of Mme Rosémilly). Male display also figures as an important motif in the entirety of *Fort comme la mort* and in "Fini" (CN2, 513–518): in both of these cases, a dashing but aging man pays constant attention to his appearance in order to impress a younger woman, the daughter of a present or past mistress. Male display also plays a thematic role in "La Relique" (CN1, 589–593) as concerns the conduct of the central male character who is attempting to impress a young woman with his ability to procure a relic for her. It also is an operative motif in "Un coq chanta" (CN1, 478–482).

Finally there are the textual attempts to short-circuit the evolutionary process, all of which end in disaster. These are particularly interesting in that they demonstrate a conflict between culture and nature: when culture tries to circumvent the natural, evolutionary mechanism of sexual selection, things go wrong and characters are hurt. This is quite obvious in cases where males, proceeding from an assumed posture of control, engage in rape: consider "La Petite Roque" (CN2, 618–650), "Madame Baptiste" (CN1, 653–658) and "L'Odysée d'une fille" (CN1, 997–1003), all discussed in chapter 5 and all ending in calamity resulting from sexual abuse.[2] In "Jadis" (CN1, 181–185) a sage grandmother bemoans the corruption of love and romance by the cultured institution of marriage, and indeed, arranged marriages throughout the corpus of Maupassant's fiction tend to be disastrous: consider the central action of *Une vie* and "Mademoiselle Perle" (CN2, 669–684). The whole point of "Ma femme" (CN1, 659–665) is that marriage without love is a lottery in which very few are lucky. For the rest, impulsiveness in sexual conduct is apparently selected against, as occurs in "Ce cochon de Morin" (CN1, 641–652, discussed in chapter 5) and in "Mots d'amour" (CN1, 358–361) where female sexual aggressiveness is advised against in an epistolary exchange.

The argument appears to be made implicitly in the corpus that the evolutionary order of sexual selection is the preferable one. Again, we are not considering the motif of marital infidelity in this discussion, for reasons of excluding the dynamics of social, legal institution from a study of the natural function of reproduction and its mate-selection process. But in any case, the Darwinian view of selection does appear to hold sway in thematics.

Heredity and parentage

No topic is more central to Darwinian theory than heredity, for it is the leaving of genetic heritage that renders evolution and the perpetuation of the species possible. A Darwinian reading of the motif would require evidence of a basic awareness of the inheritance of traits, but one which is predicated nonetheless on the genetic uniqueness of the individual within the species—variability in similarity. There is no reason to assume that Maupassant was any better informed than Darwin on the findings of Mendel, so we need not look for a literary *mise en scène* of particulate genetics and combinant traits in heredity; Darwin was aware of the results of the heredity process, even if he was not able to explain the process itself, and this is about where Maupassant himself appears to be on the issue.

In *Bel-Ami,* Norbert de Varenne, the resident intellectual on the staff of the newspaper that employs the central character, gives a brief and clear basis for all of the literary treatment of biological heredity in Maupassant. He tells Duroy,

> ...il naîtra des millions, des milliards d'êtres qui auront dans quelques centimètres carrés un nez, des yeux, un front, des joues et une bouche comme moi, sans que jamais je revienne, moi, sans que jamais même

> quelque chose de moi reconnaissable reparaisse dans ces créatures innombrables et différentes, indéfiniment différentes, bien que pareilles à peu près. (R, 300)

The basics of the evolutionary theory of heredity could find no more precise and clear expression if Darwin penned them himself. The individual is first a member of a species, and hence bears biological resemblance to others in that group. Still, as a function of biological process, heredity involves the shuffling of traits in the course of their being passed on. Traits are inherited, but the offspring may inherit them in adapted form and in unforeseen combination. The final result is that the individual within the species remains unique, and the reproductive process is the vehicle of variation (aside from chemical or radiological interference from the environment or the sort of embryological aberration induced in "La Mère aux monstres," already discussed).

Numerous characters in the corpus inherit traits from their parents, and given Maupassant's frequent pessimism, it is usually not for the best. For example, in *Une vie,* the protagonist's son Paul inherits much of his father Julien's inconsiderateness and disregard for Jeanne, and he treats her accordingly. In his adult years, he neglects his mother and contacts her only when he finds himself in material need. The licentiousness and hardness of the father emerge in the son as he reaches his majority:

> Le lendemain on retrouva le jeune homme chez une fille entretenue de la ville. Son grand-père et sa mère l'emmenèrent aux Peuples sans qu'un mot fût échangé entre eux tout le long de la route. Jeanne pleurait, la figure dans son mouchoir. Paul regardait la campagne d'un air indifférent. (R, 158)

It must be said, too, that the half-brother of Paul, Denis Lecoq who is the son of Julien by Rosalie, the maid, shows none of the undesirable traits of his father's genetic legacy. He resembles Julien physically, but seems to have his disposition from his mother's side:

> Il était rouge, vigoureux, avec les cheveux blonds et les yeux bleus de sa mère. Et pourtant il ressemblait à Julien. En quoi? Par quoi? Elle ne le savait pas trop, mais il avait quelque chose de lui dans l'ensemble de la physionomie. (R, 169)

The character's eye for traits of heredity, reported in the narrator's vision, is noteworthy. Heredity of traits, both physical and of character, is a function of physical genetics, as the traits of inheritance are visible in any case. This view of heredity is something Maupassant shares with Zola and the other Naturalists: traits are passed from generation to generation in combinant fashion, and by the luck of the genetic draw, traits of heredity are determinant in the development of character.

The short story "Un portrait" (CN2, 1051–1055) is another illustration of this point. Here, the narrator meets a famous, aging Don Juan named Milial, at whose home the narrator observes and studies a portrait of a woman. The portrait intrigues the narrator who finds out that it is the likeness of the host's mother. Near the beginning of the text, the narrator reports having questioned some women as to the particular attractiveness of Milial, and for an answer he got, "Je ne sais pas… c'est du charme." (CN2, 1051) In the portrait at Milial's house, the narrator finds a quality in the woman's eyes that evokes in his memory a verse of Baudelaire: "Et tes yeux attirants comme ceux d'un portrait", and he admits that the eyes did attract him irresistibly. He records his sensation:

> Ces yeux, ces yeux créés par quelques coups de pinceau, cachaient en eux le mystère de ce qui semble être et n'existe pas, de ce qui peut apparaître en un regard de femme, de ce qui fait germer l'amour en nous. (CN2, 1054)

Upon learning from his host that the portrait is of his mother, the narrator comes to a realization: "Et je compris alors d'où venait l'inexplicable séduction de cet homme!" (CN2, 1055) The narrative comes to a close on this observation about heredity.

The "inexplicable seductiveness" is a quality that came to Milial by heredity, and furthermore it is visible—if ineffable—to the narrator; the trait itself is a physical one and is part of the physical genetic heritage passed to Milial by his mother.

We have already had occasion to discuss the short story "Une veuve" (CN1, 533–538) in which a woman, now old, narrates events from her youth that led her to live her life as a widow: the sad story of a 13 year-old boy who, for love of her when she was 18, committed suicide. It is the boy's genetic predisposition to suicide for love that interests us here. The woman, "une vieille tante restée fille" (CN1, 533) describes the de Santèze family by way of situating her narrative:

> Oh! c'était une race singulière, des fous, si l'on veut, mais des fous charmants, des fous par amour. Tous, de père en fils, avaient des passions

> violentes, de grands élans de tout leur être qui les poussaient aux choses les plus exaltées... Rien qu'à les voir, on le devinait. Ils avaient tous les cheveux bouclees, bas sur le front, la barbe frisée, et des yeux larges, dont le rayon entrait dans vous, et vous troublait sans qu'on sût pourquoi. (CN1, 534)

This passage is very revealing. First, it situates the phenomenon in question, the lovelorn impulsiveness of the de Santèze men, explicitly as a function of heredity; then it gives a physical aspect to the phenomenon, associating the character trait with the physical traits of heredity—hair type and eye size—and then going on to link the gaze of those eyes to the rest of the inherited traits. The overall message is that the genetic legacy, looks and behavioral tendency together, is passed from generation to generation as a physical component of biology. Again, the legacy is visible, as the woman says, "Rien qu'à les voir, on le devinait."

Other texts lend themselves to Darwinian interpretation at an almost biochemical level where the issue of heredity is concerned. One such work is "Mouche" (CN2, 1169–1178) in which a fraternity of five young *canotiers* take in "une petite créature fluette, vive, sautillante, blagueuse et pleine de drôlerie" (CN2, 1171) as a combination of coxswain, comedian and mascot. One by one, the five young men begin a physical relationship with the young woman they have nicknamed Mouche. Before long, she announces that she in pregnant, and naturally the circle of male friends concludes that it would be impossible to establish the identity of the biological father. The narrator refers to "la cruelle farce de la nature qui ne permet jamais à un homme de savoir d'une façon certaine s'il est le père de son enfant" (CN2, 1175), this in an age when DNA profiling was not yet available for the precise determination of paternity as it is today. At any rate, the five friends forge a solemn pact: they agree to adopt Mouche's baby as a group and, even if paternity later becomes evident by physical resemblance, never to make an issue of it or to dissolve the pact of adoption. As it happens, one day during her pregnancy, Mouche takes a fall while getting out of the boat and suffers a miscarriage. Mouche loses touch with the group which itself eventually disbands, and the narrator recounts the whole affair as a distant memory of his youth. It is the question of paternity and the adoption pact that are of interest to us here as we consider the topic of genetic legacy.

The pact made by the five *canotiers* is something of a puzzle. Why indeed would five men, certain that four of their number cannot be the father of the coming child, pledge to adopt and support mother and child? The young man, nicknamed N'a-qu'un-oeil, who announces Mouche's pregnancy to the group, says in one breath, "Ce n'est pas le moment de l'abandonner et la recherche de la paternité est interdite." (CN2, 1175) On the very human level, one might impute a motive of sheer magnanimity on the part of the group, for whom the character Tomahawk says "Ma foi, tant pis, l'union fait force." So, unlikely as it may seem, this band of friends makes a decision not to abandon a young woman and her baby in distress, and their gesture may be understood as one of fine altruism. But to phrase the quandary in biological terms, why should four males who are not the child's father commit their resources to fostering the offspring of another? What Darwinian sense does it make to contribute to the survival of offspring when the odds are four to one that it is the child of another?

Darwinian biologists like Richard Dawkins have grappled with this very question, the evolutionary sense of altruism in a world of hostile environments and limited resources. The answer they offer is an elegant one, and it has to do ultimately with Dawkins' concept of the "selfish gene," that is, the determination of organisms' behavior at the genetic level, as a function of the genes' programmed imperative to survive and replicate themselves. In the case of "Mouche" we have five males, one of whom definitely is the biological father of the coming offspring, regardless of the fact that the four others are not. Certainly it is in the biological interest of the one to foster his genetic line. But if indeed "La recherche de la paternité est interdite," would it make biological sense to abandon the baby? The answer is no, for two reasons. First, it can be argued that a one-in-five chance of genetic replication is better than none, so there is some interest of probability to protect. Second, and more importantly, the altruistic behavior serves not so much the interest of an individual as the interest of the species: all things considered, these five men (each with his own possible 20% probability of personal genetic success in the balance) collaborate to protect the interest of a child who is certainly the offspring of one of them. In the end, altruism serves the genetic interest of the group as a whole.

The text "Le Père Amable" (CN2, 731–751) gives another twist to the motif of genetic succession. This is the tale of an old Norman farmer, Amable Houlbrèque, whose son Césaire falls in love with a young woman named Céleste. The problem is that Céleste has a young son Victor from a previous relationship. Amable persistently refuses to consent to the marriage of his son with Céleste because he cannot condone the contribution of family resources to the fostering of another man's son. Césaire, however, stands by his desire to wed Céleste, and he gets the local priest to intervene with Amable on his behalf. The priest wins Amable's begrudging permission, and after the wedding, which Amable cannot bring himself to attend, the old man is miserable with circumstances in his retirement. Césaire, working the farm on his own and eager to make his new life succeed, ends up working himself to death. Now alone on the farm with Amable and her young son, Céleste turns to the father of her child. As this man makes to take Césaire's place on the farm, Amable hangs himself.

Amable's attitude lends itself to Darwinian interpretation. His values are driven by a dedication to the interests of his own genetic line. His son proposes to subvert those interests by committing resources of survival to the son of another man. The child means nothing genetically to Amable, and he represents only a drain on resources that Amable would obviously prefer to reserve for his own line of genetic descent. Even if Céleste were later to provide offspring in that line, the child Victor constitutes a diversion of some means of survival that otherwise would go to the Houlbrèque lineage. It might be a different story if the Houlbrèques were wealthy, but the narrator makes a point of stating that, prior to any change in Césaire's marital status,

> ...ils vivaient péniblement, bien que Césaire fût un bon cultivateur. Mais ils ne possédaient ni assez de terres, ni assez de bétail pour gagner plus que l'indispensable. (CN2, 736)

So it is clear that the economic environment puts stress on the Houlbrèques' struggle for survival, and any draw on resources will have an impact on the lineage's legacy. Obviously Amable regards Victor as a threat to the interests of the Houlbrèque line. After the marriage to which the old man consents most reluctantly, and after the death of Césaire (the last male of the family line of an age to reproduce), the seventy year-old Amable realizes that "life" is over, if life's worth is defined in the Darwinian context as leaving progeny. Seeing the end of his genetic heritage, and seeing his own material resources about to fall into the hands of the competing genetic line of Victor and his biological father, Amable commits suicide. His death is a reflection of that of his lineage.

The reader must be asking, If we apply a Darwinist reading to "Le Père Amable" and find that concern for the genetic line is the determinant motif in action, what about the short story "Le Papa de Simon" (CN1, 74–82) where we saw earlier that the adoption of a fatherless boy made a hero of Philippe Remy? Does not the Darwinian model apply to that text? Not that it necessarily has to, but the model actually does apply to "Le Papa de Simon." Consider the difference between the two situations: Amable Houlbrèque is the last surviving male of his line, and is no longer eligible to procreate, so he has absolutely nothing to gain where heredity is concerned; Philippe Remy may be dividing his resources between Simon (the offspring of another man) and his own prospective line after his marriage to Simon's mother, but he has much to gain by his choice to adopt Simon. First, by adopting the boy, he enhances his stature in the eyes of Simon's mother—he is improving his courtship display by demonstrating his physical ability and willingness to defend his new mate and her interests:

> Tu leur diras, à tes camarades, que ton papa, c'est Philippe Remy, le forgeron, et qu'il ira tirer les oreilles à tous ceux qui te feront du mal. (CN1, 82)

Then, by establishing himself as the new man in the life of Simon's mother, Philippe assures himself the opportunity to father his own genetic line. So he stands to gain considerably by dividing his resources to protect Simon. Césaire Houlbrèque finds himself in a similar situation at the outset in "Le Père Amable" and he makes the same choice to support the son of another as part of his marriage commitment; when he dies, his father Amable has no incentive (genetically speaking) to contribute resources to young Victor.

In the cases just studied, heredity and genetic succession are detailed thematically in a most positive light. Characters put high value on them, and go to some lengths to protect them—consider only Maréchal in *Pierre et Jean* who is obviously willing to compromise his former mistress, Mme Roland, by endowing Jean (their son) to the exclusion of Roland's son Pierre. In other settings, however, the motif of the genetic line is set forth as a negative force. In a way familiar to any reader of *Les Rougon-Macquart*, for example, heredity can be presented as "chargée" or laden with a past that bodes ill for the future. Such is the case in some of Maupassant's fiction.

In the well-known short story "Garçon, un bock!..." (CN1, 1123–1129), the narrator recounts his chance meeting with an old schoolmate who, now in adult life, spends all his days and evenings in a café ordering beer. This old acquaintance, Jean des Barrets, tells his sad story to the principal narrator by way of explaining his present state. As a boy, Jean happened secretly upon a scene of confrontation between his parents, in which his father was trying to wrest his wife's financial resources from her, having squandered all his own on a profligate lifestyle. Jean's mother tenaciously refused to give in, insisting that she was protecting young Jean's inheritance. Jean's enraged father struck the mother in the heat of the argument, and the boy remained traumatized all his life by the incident. He says now to his friend,

> Eh bien, mon cher, c'était fini pour moi. J'avais vu l'autre face des choses, la mauvaise; ... Mais je n'ai plus eu de goût pour rien, envie de rien, d'amour pour personne, de désir quelconque, d'ambition ou d'espérance. (CN1, 1128)

The material heritage that his mother had guarded for him at the cost of her physical safety was what sustained him now in his lethargic, unproductive life. Obviously a measure of his sadness comes from the fact that what his mother had risked her life for ended up serving no good purpose.

What does all this say about heredity? First, Jean's father evidently passes on to his son a tendency squander resources in an unproductive life. Granted, Jean does not indulge in excess as his father had, but his life in the end does not amount to any more than his father's; all he has to show for his mother's fortune is a life of consuming beer and tobacco. Then there is the notion that the sins of the father become a burden on the son, one that he bears to the exclusion of any other enterprise in life. The trauma of having witnessed the violent argument between his parents stays with Jean for the rest of his days, influencing his entire life's experience. And this to the point where his mother's material legacy is itself consumed in waste.

There is a similar outcome in *Pierre et Jean* in which Pierre is so scandalized by the revelation of his mother's marital infidelity that his opinion of women in general is forever jaded and he goes into a tailspin that affects his career. While his fair-haired half-brother Jean appears to have inherited the lighthearted insouciance of his biological father, Maréchal, the dark-complexioned Pierre becomes a creature of the night who goes on thinking dark thoughts about women and life on the whole. The whole experience of learning of his mother's past, seeing his father's obliviousness, and watching his brother gain a fortune and a marriage, leaves Pierre quite sour on life; but he also inherits a degree of his mother's guilt and his father's mediocrity, which leads him to despair of life and to go into a kind of exile in the end. The mother's "crime," while repugnant to Pierre as he repeatedly thinks, also constitutes for him a stigma on the family and is thus to him a mark of collective guilt. The family secret is enough to change his life for the worse and drive him away. The half-brother Jean, for his part, is glad enough to be free of part of the Roland family heredity:

> Il lui en coûtait moins d'être le fils d'un autre; et après la grande secousse d'émotion de la veille, s'il n'avait pas eu le contre-coup de révolte, d'indignation et de colère redouté par Mme Roland, c'est que depuis bien longtemps il souffrait inconsciemment de se sentir l'enfant de ce lourdaud bonasse. (R, 817)

The physical signs of heredity, and the detection and observation of these signs, constitute a motif of importance in Maupassant. Again, at a time prior to the availability of blood typing or of DNA profiling, characters in these texts rely on the study of physiognomy, coloration, body type, and so on in their attempts to draw conclusions about parentage. Their efforts pass for scientific. The most graphic case of this is in *Pierre et Jean* where the physician Pierre spends much of his energy in the narrative studying his brother's face and physical traits, trying to see in him the heredity of Maréchal, the man whose death leaves the brother wealthy, and Pierre penniless. Pierre first sneaks into Jean's room at night to observe his features closely, looking for some visible sign of Maréchal's paternity, and he is motivated by a physician's attention to biological heredity: thinking of the relationship between Jean and Roland, the man everyone takes for Jean's father, Pierre's thoughts say,

> Aucun lien, même le plus léger, ne les unissait, et ils ne le savaient pas! Ils se parlaient avec tendresse...comme si le même sang eût coulé dans leurs veines. Et deux personnes nées aux deux extrémités du monde ne pouvaient pas être plus étrangères l'une à l'autre que ce père et ce fils. (R, 771)

But out of worry that he is jumping to conclusions, Pierre sets about studying his brother physically:

> Comment le savoir? Ah! si une ressemblance, même légère, pouvait exister entre son père et Jean, une de ces ressemblances mystérieuses qui vont de l'aïeul aux arrière-petits-fils, montrant que toute une race descend directement du même baiser. Il aurait fallu si peu de chose, à lui médecin, pour reconnaître cela, la forme de la mâchoire, la courbure du

> nez, l'écartement des yeux, la nature des dents ou des poils, moins encore, un geste, une habitude, une manière d'être, un goût transmis, un signe quelconque, bien caractéristique pour un oeil exercé. (R, 771)

Seeing no resemblance between his sleeping brother and his own father, Roland, Pierre determines that he must get hold of a small portrait of Maréchal that used to be on display in their house. He catches his mother in a lie about whether the portrait is still in their possession (R, 774 and 779), and upon studying the image and comparing the faces of Jean and Maréchal, he concludes:

> Elles avaient, certes, des signes communs: la même barbe et le même front, mais rien d'assez précis pour permettre de déclarer : "Voilà le père, et voilà le fils." C'était plutôt un air de famille, une parenté de physionomies qu'anime le même sang. (R, 780)

Thus the motif of physical traits of heredity takes center stage in the novel. The same motif arises in similar circumstances in the text "Monsieur Parent" (CN2, 580–617) where the central character is seized with doubt as to his paternity of the child Georges he and his wife have been raising and to whom he has consecrated his whole life; he suspects that a friend, Limousin, is the real father, on the say-so of the maid:

> Alors, peu à peu, s'éveilla dans son âme un désir affolé de regarder Limousin assis en face de lui et qui roulait des boulettes de pain. Il voulait voir s'il ressemblait à Georges. ... De seconde en seconde, il jetait un coup d'oeil rapide sur ce visage, cherchant à en reconnaître les moindres lignes, les moindres traits, les miondres sens; puis, aussitôt, il regardait son fils, ... (CN2, 595).

Heredity is of course the central thread of interest in a Darwinist model of study, and it leads directly to the general motif of parentage as we pursue this reading of the fiction of Maupassant. Many of the stories and novels revolve around the question of parentage rightly or wrongly assumed. Often there is the figure of the man who, having abandoned the offspring of his youthful amorous adventures, seeks for one reason or another to renew his acquaintance with his progeny. We have already seen this, for example, in "Un fils" (CN1, 416–425) where a member of the Académie française relates his yearly trip to Brittany to see the son who grew up to be rather unproductive and unlikable through his father's neglect. The academician scrutinizes the younger man's face

> ...pour décider si, oui ou non, nous avions des traits communs.
> ...et je lui donnai cent sous en le dévisageant anxieusement.
> (CN1, 423)

In his narrative, the academician clearly implies his regret and his sense that parentage should involve more than the biological transmission of life and genes. The Darwinian reading of this development is that parenting can, by choice, involve the fostering of offspring which enhances the latter's chances of survival and success, thus enhancing the genetic line's chances of perpetuation. This is, after all, the man who hears his senator friend speak of the ebony tree's insouciance at spreading its seeds to the four winds, and who replies, "Nous en faisons autant, mon ami." (CN1, 416–417), and goes on to preface his story with the remark, "Tenez, j'ai, pour ma part, sur la conscience une très vilaine histoire..." (CN1, 418). The human emotional bond between parent and child has justification in the interpretive model of evolutionary theory.

In numerous texts in the corpus we find permutations of the motif of parentage neglected. The short story entitled "Le Père" (CN1, 1071–1079) is the narrative of the eponymous character who, having abandoned the woman he impregnated, feels remorse ten years later upon seeing the woman with a ten year-old son; the father makes contact with the woman and her husband, and asks only to embrace the boy. After so doing, he goes away miserable.

"L'Abandonné" (CN2, 225–233) opens during the vacation trip to Normandy of an elderly married couple and their "friend." One afternoon, the woman and the friend go off for a walk, but their real destination is the home of their adult son whom they abandoned in his infancy and whom they hope to see for the first time. They meet briefly with him and leave very sadly before revealing their secret.

"Duchoux" (CN2, 996–1003) shows a father who is more cynical. Here an aging Parisian socialite gets the urge to track down his grown bastard son in Marseille and sets out to visit him. Upon arriving, still unknown to his son whom he had supported anonymously over the years, he pretends to have business with him and observes the sights and smells of family life in his household. The father is put off by his grandchildren and everything he senses in the house, so he leaves before revealing his identity, and resumes his sterile life in Paris.

"Le Champ d'oliviers" (CN2, 1179–1204) presents a southern priest who one day is confronted by an unsavory young man who is his son by a relationship maintained before his entry into the clergy. The son reveals to the priest how his mother had tricked him into believing that the priest's former rival was the father, and tells him of the sordid life he has led in the twenty-five years prior to the present. The priest, horrified, kills himself with a knife and the son is arrested for his murder. If ever there were Maupassant tales where parentage serves as a destructive force, this is one of them. The neglect of the father, albeit unintended, comes back to cause the destruction of the parent and the offspring as well.

"Un parricide" (CN1, 553–559) is another such tale. Here the two parents who had abandoned an infant born out of wedlock attempt to contact their grown son and to endow him without revealing their relationship to him. The couple's attempt at secrecy backfires, for the young man guesses that these are his long-lost parents. He confronts them, a violent scene ensues and the man ends up killing the older couple in self-defense. In this, as in the other narratives regarding the parentage motif, the descendant relationship, if it is nothing else, is a catalyst for precipitous action in Maupassant's fictional world. The genetic line is charged with meaning for characters, and its implications go far beyond the mere transmission of life to the more profound issues of identity and heredity.

Parentage is the central motif in other texts involving infant children and the decision to snuff out the genetic line. In one case, "La Confession" (CN2, 371–378), an upstanding and influential man leaves a confession in his will, admitting to having exposed his infant son (by a young mistress) to the cold and bringing about his death. Here, the crime is provoked by the man's attention to the "purity" of his genetic legacy: he preferred to leave children by a wife chosen for her social acceptability, rather than by a mistress. In "Rosalie Prudent" (CN2, 699–702), a young unmarried woman is acquitted, out of pity, of charges in the killing of her infant twins, in recognition of the fact that she could neither support two children nor choose one to live and one to die. For this text, the application of the model of evolutionary heredity frankly must cede to the Social Darwinist model for its attention to the harsh realities of the economic environment.

In other texts, however, the more properly Darwinian reading may be suggested metaphorically by the development of the motif of parentage and heredity in the thematics of the piece. In "Le Testament" (CN1, 620–624), a young man discovers in a will that his biological father was actually his mother's lover. He makes the decision to split his financial inheritance with his half-brothers, but to take the name of his true father. We may interpret the material inheritance figuratively as means of survival which the protagonist shares with those who also share at least in half of his genetic lineage.

And of course, with Maupassant, the motif of parentage can just as well be a pretext for humor. In "Le Cas de Mme Luneau" (CN1, 963–967), a plaintiff is suing Mme Luneau for payment of 100 francs, promised in the event that the man impregnated her; Mme Luneau produces numerous witnesses who attest that the baby may well be theirs. Seeing the impossibility of a definitively accurate ruling, the judge awards a compromise of twenty-five francs to the plaintiff. This text we may note as an example of the bartering of parentage, with no particular interest in the hereditary aspects of the matter.

One variation on the motif of parentage, that of parentage falsely assumed, is another great catalyst for action in Maupassant's fiction. Of course, the entire novel of *Pierre et Jean* revolves around this thematic. Pierre the physician becomes Pierre the detective, building the case against his mother for a past infidelity that resulted in the birth of Jean. When Jean alone inherits the fortune of an old family friend, Pierre's discovery of the truth of Jean's parentage leads him to reexamine his beliefs and his values concerning family, femininity, and even himself. In a passage cited above, Pierre questions the relationship between Jean and Roland, the man who has always passed for Jean's father, and their genetic remoteness now disturbs him. The discovery of the truth drives a wedge between the brothers which is genetic in nature: Pierre is now conscious of Jean's Maréchal half and of the difference between that and the Roland genetic legacy (which Pierre finds to be undistinguished, to say the least). In this way he comes to call his own self-worth into question. His own heredity is from a woman he judges as fallen and a man he finds mediocre. If heredity is the power of the past to determine the future, Pierre does not see much of a life ahead for himself.

"Monsieur Parent" and "Le Testament" discussed just above raise similar implications for the characters who belatedly discover the truth of their parentage. In "Monsieur Parent," the last the reader sees of young Georges is that he is "éperdu" and leaning against a tree in the shock of hearing the identity of his father called into question (CN2, 616). The first-person narrator of "Le Testament" refers to his mother's husband as "un de ces rustres qu'on appelle des gentilshommes campagnards" before detailing the man's philandering and his cruelty to the narrator's mother. The implicit narratorial tone is one of relief to be free of that man's heredity.

"Le Petit" (CN1, 957–962) sets a context by now largely familiar. Here a man has lost his wife in childbirth, and he raises the child lovingly with the constant companionship of another man who turns out to be the child's biological father. In a dispute with the maid over the child's diet, the widower is informed of the truth by the angry maid who, incidentally, appeals to physical traits of resemblance to make her point:

> Regardez-le, regardez-le donc, bête que vous êtes, si ce n'est pas tout le portrait de M. Duretour; mais regardez son nez et ses yeux, les avez-vous comme ça, les yeux? et le nez? et les cheveux? (CN1, 961)

The truth of the situation leads the widower to commit suicide and to leave a note confiding the child's care to the biological father. As usually happens in Maupassant's narratives centering around this motif of identity and parentage, the truth about genetic heredity leads to a redefinition of relationships of dependency and a reassessment of self-worth.

Reproduction as evolutionary success

The mark of success in the evolutionary struggle of a species is obviously the strength of its numbers from one generation to another. By a Darwinian model of study, the mark of an individual's success set against the broad background of evolution by natural selection is the same, the passing on of one's genes to a subsequent generation. This signals both the perpetuation of the genetic line itself as well as the opportunity to pass on genetically the advantageous adaptations that helped the individual to get ahead enough in the struggle for life to reproduce. Advantageous adaptations are of no worth to any other than the individual unless they are passed along by heredity to the genetic line. In this way, reproduction is the vehicle of evolution as much as heredity is the evolutionary "leg up" in the struggle for survival.

Maupassant's fiction shows an awareness of the importance of leaving progeny both for the good of the species (or social unit, as the case may be) and for the legacy of the individual. The main motif of "La Reine Hortense" (CN1, 802–809) is the very thing in question. Hortense has spent her life unmarried but exercising firm economic control over her entire family. As she lies on her deathbed, she is attended by others in the family whose numerous children playing about the house strike a contrast with Hortense's personal solitude. Hortense begins to have delusions, speaking out loud to imaginary sons and daughters:

> Viens ici, mon petit Philippe, embrasse ta mère. Tu l'aimes bien, ta maman, dis, mon enfant? Toi, Rose, tu vas veiller sur ta petite soeur pendant que je serai sortie...
> ...Henriette! ...Dis à ton père de venir me parler avant d'aller à son bureau... Je vais te faire pour le dîner un plat de riz au sucre. Les petits aiment beaucoup cela. C'est Claire qui sera contente! (CN1, 805)

With the invention of make-believe offspring, the implications are clear: Hortense, despite being the "queen" of the family, has left no progeny and feels the void as she is about to die. To whatever extent her life has been important and effective, from an evolutionary perspective there is one success that has eluded her.

There are elsewhere two female characters whose textual identity is defined for the most part by the fact that they have given no continuity to their genetic line. The figure of tante Lison in *Une vie* is qualified in the narrative (that is, in the narrator's reportage of the family's collective perception of the woman) as one who "n'avait jamais compté pour rien dans la famille", and in her own eyes, she is "inutile et importune" (R, 36). Of her it is said,

> Sa soeur, par habitude, prise dans la maison paternelle, la considérait comme un être manqué, tout à fait insignifiant. (R, 37)

In the view of the "maison paternelle," Lison has not led a life fulfilled. The term "être manqué" defines her by what is missing in her life. The scene of Lison's sudden and unexpected breakdown one evening before the wedding of the novel's central character points up the sterility of her existence (R, 40).

This scene and its principal character of the unwed aunt are abstracted for the short story "Par un soir de printemps" (CN1, 309–313). Under the same circumstances of an impending wedding of her niece, the aunt (again named Lison) has a breakdown to the surprise of the family members who afford her no attention, and the breakdown is obviously a response to the realization that her life lacks substance. The narrator says of her flatly, "Elle ne tenait point de place" (CN1, 311). Substance in this culture of "maison paternelle" is ultimately a contribution to the genetic line. The phrase recurs, "Elle...ne comptait presque pas dans la famille" (CN1, 310). Implicitly, "family" means succession and continuity of the line, and one who leaves no offspring loses importance.

The implicit judgment is not reserved for women by any means. The entire inserted narrative in the short story "Après" (CN2, 1246–

1251) is an apologia in the first person told by an aging priest who explains that he decided young to forgo the fulfillment of fatherhood in order to spare himself the grief of possibly losing a loved one. When his hostess's small children have gone off to bed, he states his case in response to the charge, "...dites-moi comment vous vous êtes décidé à renoncer à tout ce qui nous fait aimer la vie, nous autres, a tout ce qui nous console et nous soutient" (CN2, 1246). What sustains us and makes life worth living, by the accepted values in the conversation, is the presence of children. This sentiment may be in concert with the biblical dictum to "be fruitful and multiply," but it is by the same token a way of underscoring the evolutionary imperative to contribute to the continuation of the genetic line.

Adoption

A Darwinian rationale has already been offered for the decision to adopt and foster the offspring of another's line, in the case of "Le Papa de Simon." The motif arises in other texts as well. In "L'Enfant" (CN1, 483–488) a young man marries and learns on his wedding night that his former mistress is about to die in labor giving birth to his child. He rushes to her side and grants her dying wish to swear not to abandon the child. He returns with the baby to his bride who generously proposes that they raise him together. The impetus to adopt in this case is obvious: while it may not be the offspring of a marital union, the child does represent the man's own genetic line.

Another case is closer in motif to that of "Le Papa de Simon." In "Histoire d'une fille de ferme" (CN1, 225–243) an unwed mother is forced to give her child up, but does all in her power to stay in contact with him. Later she is coerced into marrying her employer who, when he realizes that he is sterile, agrees to adopt his wife's son. His gesture can be ascribed to one or both of the following biological motives: either he is trying to enhance his attractiveness in his wife's eyes (as a function of sexual courtship, in the manner seen with Philippe Remy in "Le Papa de Simon"), or he accepts surrogate parenthood to replace genetic succession when the latter is not possible. Since he makes the offer to adopt the boy only after accepting the fact that he himself is sterile, it seems that the second rationale is more plausible than the first.

The aim of this chapter has been to demonstrate the dynamism as well as the breadth of the elements of heredity and sexual selection in a Darwinian reading of Maupassant. Clearly these issues have merited separate study since, while they stand apart from the natural selection that is the cornerstone of Darwin's theory, they do represent forces which act together with natural selection to produce the phenomenon of evolution. Our survey has given evidence of the fact that Maupassant wrote these Darwinian motifs into much of his fiction, and that the motifs play out in ways that are in keeping with the logic of evolutionary theory. Sex drives appear as extensions of nature, and this in spite of any pressure that a cultured system of morality can mount. Biological principles of reproduction, including a genetically reasonable taboo against incest, serve as bases for the literary expression of motif that is enlightened by evolutionary theory. Sexual selection, rivalry and lures all function in ways that a Darwinian model might predict. Heredity and parentage are nothing short of central to Maupassant's thematics—and this comes as no surprise to anyone familiar with the literature of the French Naturalists and their scientistic bent, a slant which Maupassant retained from his early contacts with the school.

Having discussed those points that make up the core of the Darwinian model—the nature of species, natural selection and the struggle for life, sexual selection and heredity—we turn finally to remaining miscellaneous topics pertinent to evolutionary theory which relate to the corpus of Maupassant's fiction.

NOTES

1. I refer to Maupassant's career as an author of novels and short stories, for we cannot completely overlook his collaboration in "une pièce absolument lubrique," *A la feuille de la rose, maison turque* in April, 1875, a bawdy farce composed for his friends' enjoyment.
2. See the work of Sharon P. Johnson, 135–175, on this topic.

Chapter 7: Conditions, Environment and Population

The environment is the crucible in which evolutionary theory sees the struggle for survival played out. The forces and conditions which the earth amasses blindly are those which shape the destiny of individuals and species. Characteristics of this earth that came as news to the latter half of the nineteenth century were its great age and, most of all, its unrelenting amorality and insentient lack of concern. The slow pace of accumulative evolutionary change was shown by Darwin to be predicated on a new geological sense of time, one that made of the earth a planet far older than preceding cosmology had imagined. Any subsequent world view that was to be at all in harmony with evolutionary thought would necessarily accept the revolutionary perspective on "deep time" and an earth whose surface features have been transitory for billions of years. No approximation of Darwinist perspective can be complete or consistent without a geology that supports it.

Maupassant's attention to descriptive detail in setting the environment of his texts' fictional struggle for life would seem to bear witness enough. In all that has gone before, we have seen a depiction of the physical world in Maupassant that is very much in keeping with the evolutionary view of life. But what of geology in particular? Does Maupassant's fiction ever attest to awareness of things geological and of the age and dynamism of the earth's surface? We may cite three examples of such a motif, all in texts set in the Auvergne region of central France. "Malades et médecins" describes the surface;

> Tous ses volcans éteints semblent des chaudières fermées où chauffent encore, dans le ventre du sol, des eaux minérales de toute nature. De ces grandes marmites cachées partent des sources chaudes.... (CN2, 101)

The volcanoes exercise an attraction for Maupassant's narrators and for characters who make a point of going out to tour them. Like the diarist in "Mes vingt-cinq jours" who is delighted to make a long excursion in order to visit "un admirable petit lac...gîté dans le fond d'un ancien cratère" (CN2, 535), the narrator of "Malades et médecins" is taken with the ancient past and the prehistoric activity of the volcanic craters.

The action of the novel *Mont-Oriol* is also set in Auvergne and is put in motion with the motif of geology (i.e. the distribution and flow of thermal springs) with an engineer among the characters to explain early on the finer points of water flow (R, 511). The narrative makes many mentions of the craters in the region, as well as of their ancient past. The character Paul Brétigny recounts to the central character Christiane his travel experiences in Sicily where he was very moved to stand "au bord du cratère monstrueux de l'Etna" and to be "penché sur cette bouche effroyable de la terre" (R, 558).

Materialism

The world is, for Maupassant in his fiction, a material environment where principles governing physical matter tend to hold sway. There is consistency in Maupassant's world view, for the anti-dualism of Schopenhauer goes hand in hand with the ideological materialism of the Darwinian view of life. In general, the forces which move character and action in Maupassant are physical (or not far removed), and the realities of the environment are themselves brute forces. The passions find expression in the physical world, and it is easily understandable that volcanoes, for example, can be symbolic of those passions as they are in *Mont-Oriol*. The materialism of Maupassant's world catches the eye in the absence of any spiritual or religious protocol; the many of his texts in which the *fantastique* comes into play tend to bring eerie phenomena into the material world. The "Horla" moves objects, casts a shadow in the mirror, and meets a physical end in fire; other textual doubles are spoken of by narrators as entering into the physical body.

In a very real sense, Maupassant was abreast of his times where the textual rendition of ideological materialism is concerned. The thinkers who had the greatest currency in the latter half of the nineteenth century were those who held to it. Marx's model of historical interpretation was one of dialectical materialism. Even Freud with his notion of the unconscious issued a challenge to the model of Cartesian dualism which identified the mind with consciousness; Freud instead made the body and the history of the subject count for much in psychoanalysis. Schopenhauer was the one thinker whom Maupassant read as a young man. And of course, there was Darwin.

Darwin's brand of materialism has to do with ascribing life, and all its variegated species and its countless accumulated changes, to material causes and processes. He looks no further than the material world for explanations of life and evolution, first, because he is a scientist and hence he deals only in the empirically verifiable and understands life as a physical phenomenon, and second, because there is sufficient cause to be observed in the physical world for the testing of his hypotheses (given the "deep time" that geology now admits, and given the rigors and the effectiveness of nature's inexorable winnowing process which he dubbed natural selection). Evolutionary theory has not, since Darwin, looked to metaphysical sources of explanation because the physical world has provided answers to all of its questions and evidence in abundance to sustain its interpretations and conclusions.

It would be wise here to reiterate the caveat raised in chapter 2, that we should not seek to make Maupassant appear to be "a Darwinist." It has not been a goal of this study to label him thus, nor should it be, according to the evidence; the occurrence of motifs of evolutionary theory in Maupassant are definitely not of a rhetorical sort, nor does he ever pose as an apologist or even as an exponent of Darwinian ideology as such. But it does appear to stand that Maupassant, in infusing many of his narratives with motifs borrowed from Darwinian theory (as the preceding four chapters have labored to demonstrate), also adopts premises and assumptions of Darwinian world view. One such premise is the sufficiency of material causes for physical phenomena and the precedence afforded in his thought to the material. In all of the stories and novels, rarely is conflict of the ideological sort (as it is to a slight extent, for an exceptional example, in "L'Ami Joseph," CN1, 836–841). Instead, characters' needs tend to relate to either the struggle to survive or their love interests (which may be viewed through the Darwinian lens as a function of the drive to reproduce). In either case, and this is the point, the conflict seeks resolution in the physical world. The *nouvelle* often cited above, "Le Vagabond," is a perfect illustration of this generality. The vagabond's conflict is to stay alive, and adversity comes in the form of physical hardship and deprivation. He starts out as a decent enough person before experiencing life on his own, but his treatment at the hands of the physical world (economic shortfall, lack of shelter and food, isolation, etc.) overcomes any social sense of morality in him: justice becomes whatever it takes to meet his material needs.

In general, the reduction of the abstract to the concrete in Maupassant is one of the traits that do link him to the Naturalists as a group. An aspect of their shared vision and technique is to focus on the conflicts of the physical world and to eschew the traditional spiritual consolations for those material crises. Maupassant, for his part, tends to do just that. Few of his *vagabonds* turn to the Church under any circumstances, and none turns to ideology of any sort. Confession of faith is either pro forma or motivated by some tangible interest (as in "La Confession de Théodule Sabot," CN1, 1021–1029), or else it is transitory, as for Christiane in *Une vie*. Crises of conscience and the salvation of the soul are not of interest as Maupassant shapes his novelistic world. "Saint" Antoine (CN1, 772–779) is no saint at all, and he loses no sleep at all over his wrongdoing. The murderous and pathologically lustful mayor in "La Petite Roque" (CN2, 618–650), even if he is driven by guilt to suicide, seems to make his final expiation when he ends his earthly life. Evil is defined much less as a moral issue than as a function of insanity (witness the incidence of the motif of madness or the number of text titles containing the word "fou") or as a by-product of the struggle for survival.

"Suicidés" (CN1, 175–180) is a text which, in its thematics of *angst* in a godless universe, establishes a thematic link between Maupassant and the chain of thinkers leading up to the existentialist movement of the twentieth century. Its large part poses as the transcription of the suicide note of one who finds himself alone in a life without meaning; the physical world has lost its significance for him as he revisits the material objects surrounding him, and there is no mention of any metaphysical dimension of life to give spiritual meaning of any kind. The writer of the note, known only as Robert, expresses a feeling of contingency in a universe where one is alone and where material objects have not the faculty to give meaning to life. He writes,

> Tous les événements de l'existence qui, autrefois, resplendissaient à mes yeux comme des aurores, me semblent se décolorer. La signification des choses m'est apparue dans sa réalité brutale; ...Nous sommes les jouets éternels d'illusions stupides et charmantes toujours renouvelées. (CN1, 176)

But most revealingly, in a passage that in a way presages a Sartrian sort of *nausée,* even if in reverse causation, Robert links his feelings of indigestion to his existential crisis, and throughout underscores the material nature of life:

> ...une bonne digestion est tout dans la vie. C'est elle qui donne l'inspiration à l'artiste, les désirs amoureux aux jeunes gens, des idées claires aux penseurs, la joie de vivre à tout le monde,... Un estomac malade pousse au scepticisme, à l'incrédulité, fait germer les songes noirs et les désirs de mort. (CN1, 177)

The matter of the *fantastique* has come up, and on the face of it, it would seem incompatible with an overall materialism in a corpus of fiction. It would be somewhat contradictory to adhere to a view of existence as a physical phenomenon, on the one hand, and to a thematic of the eerie and of possessing spirits, on the other. But first, the texts of the *fantastique* may be considered to form a discrete sub-group among themselves, so for this reason they should not be discussed along with the rest since the regular rules do not seem to apply. Simply put, the world of the *fantastique* stories is a different novelistic world from that in the rest of the corpus. Furthermore, the experience of the fantastic is most often explicitly set in the context of insanity.

The second point to be raised in this regard is that there actually are several places in the corpus where characters do set the two "worlds" or codes of belief in opposition one to the other. "Magnétisme" (CN1, 406–410) is the clearest example of the explicit opposition of the two where the materialistic viewpoint is favored. In this text, set as one of the many after-dinner group conversations in Maupassant, the secondary narrator steps up to recount two cases of apparent telepathy, only to debunk them both. His stance is one of skepticism:

> ...je ne crois pas par principe. De même que d'autres commencent par croire, je commence par douter. (CN1, 407)

But the rule is not hard and fast, and once again Maupassant presents a contradiction as if to put some distance between himself and an ideological system of any kind. The short story "La Peur" (CN2, 198–205) is a thematic mirror-image of "Magnétisme." Here the secondary narrator (riding in a train with the main narrator) recounts tales of the fearful experience of the unexplainable; in the absence of empirical explanation and of proof of the supernatural, this character prefers to give the benefit of the doubt to the latter. However, it is essential to point out that this character's disposition has to do with nostalgia for the Romantic: his preference is for a sense of the fantastic and he does not deny the inroads that science has made into the fantastic; to the contrary, he laments only the loss of mystery, saying,

> A mesure qu'on lève les voiles de l'inconnu, on dépeuple l'imagination des hommes. Vous ne trouvez pas, monsieur, que la nuit est bien vide et d'un noir bien vulgaire depuis qu'elle n'a plus d'apparitions.
>
> On se dit: "Plus de fantastique, plus de croyances étranges, tout l'inexpliquable est expliquable. Le surnaturel baisse comme un lac qu'un canal épuise; la science, de jour en jour, recule les limites du merveilleux." (CN2, 199)

What is interesting in this character's speech is that it does not refute the empirical, but expresses only regret for the loss of what was poetic in prior times. Of "notre terre" in his present time, he says, "Les croyances sont parties qui la rendaient poétique." (CN2, 199) He does nonetheless ridicule the medical interpretation of a cholera epidemic as the result of microbial infestation, and prefers to interpret the epidemic as a spirit come from the Orient. The text is careful to insulate these views in a secondary narrator (while the presence of a main first-person narrator affords the text a degree of "deniability" for retrograde anti-scientific thought) and the general code of ideological materialism is allowed by default to maintain its precedence over superstition and spiritualism.

But if the reader of Maupassant should entertain any doubt on this score of spiritual versus material, he or she need only consult the last of Maupassant's fictive writings, a novel fragment called *L'Angélus*, left unfinished when Maupassant's mental illness put an end to his career. In the last passage of the text, the narrator undertakes an attack on his perception of the deity. This is a God responsible for death and disease, particularly of the physical causes of disease:

> Il a inventé le choléra, la peste, le typhus, tous les microbes qui rongent le corps, les carnassiers qui dévorent les faibles animaux. Seules, cependant, les bêtes sont ignorantes de cette férocité, car elles ignorent cette loi de la mort qui les menace autant que nous. Le cheval...et le rossignol...ne savent pas l'éternel massacre de ce Dieu qui les a créées. (R, 1223–1224)

Other texts set up the opposition of materialism and religious belief. In the light of our topic of a Darwinist reading of Maupassant, we should be very mindful of the controversy—much publicized—that evolutionary theory sparked with adherents of religious tradition. Guy de Maupassant was not yet ten years old when Thomas Huxley debated Bishop Samuel Wilberforce at Oxford (June 30, 1860) in defense of his friend, Charles Darwin, but this was only the opening of hostilities between the two camps. All Europe and the New World were abreast of the controversy which was to spill over into the twentieth century unresolved. To express even a novelistic world view of ideological materialism was to take a position in the debate, and there can be no doubt about Maupassant's position in view of his general novelistic world view and what all of his biographers agree was a fairly anti-religious bias in the man.

In the stories where religion figures abstractly as a motif, the textual *prise de position* is consistent and clear. The first-person narrator of "Miss Harriet" (CN1, 876–895) speaks representatively as he describes Harriet's devotion to her religion:

> Oh! c'était un drôle de bonhomme que son bon Dieu, une sorte de philosophe de village, sans grands moyens et sans grande puissance, car elle se le figurait toujours désolé des injustices commises sous ses yeux—comme s'il n'avait pas pu les empêcher. (CN1, 887)

The stark contrast between the world idealized by religion and the "real, " material world points up the bias of thought in Maupassant. It is not so much a hostility towards religion as a conviction that the material plane is the only one for which there is evidence of efficacy. This is, incidentally, the same position of agnosticism which Darwin came to espouse. In any event, "Miss Harriet" is the narrative of a devout Englishwoman whose first encounter with love, her infatuation with the narrator (a young artist at the time), leads to her suicide. For this character, the abstraction of religion is ultimately unable to withstand the challenge of the earthly experience of love attraction. When she observes the narrator courting a young maid, Harriet throws herself down a well. The subsequent reflection of the narrator bears upon Harriet's beliefs and the world:

> Je sentais peser sur cette créature humaine l'éternelle injustice de l'implacable nature! C'était fini pour elle, sans que, peut-être, elle eût jamais eu ce qui soutient les plus déshérités, l'espérance d'être aimée une fois!...Pourquoi aimait-elle d'une tendresse si passionnée toutes les choses et tous les êtres vivants qui ne sont point les hommes?
>
> Et je comprenais qu'elle crût à Dieu, celle-là, et qu'elle eût espéré ailleurs la compensation de sa misère. Elle allait maintenant se

> décomposer et devenir plante à son tour. (CN1, 894)

The narrator, materialist that he is, describes Harriet's end in its physical aspects and sets aside the spiritual with a passing bow to her belief, in recognition of its ineffectiveness in the outcome of her life. In the narrator's words, it is nature that is responsible for her death, and it is to nature that Harriet returns with the decay of her body.

In other places, characters set religious belief in opposition to their own philosophies. The opinionated Joseph of "L'Ami Joseph" squares off with his host's local clergyman in the following terms:

> "Votre Dieu, monsieur, est de ceux qu'il faut respecter, mais aussi de ceux qu'il faut discuter. Le mien s'appelle Raison; il a été de tout temps l'ennemi du vôtre...". (CN1, 839)

We have already had occasion to detail the beliefs of the baron (father of the central character) in *Une vie* and his distaste for the brand of religious intolerance preached by l'abbé Tolbiac. The entire text of "La Légende du mont Saint-Michel" (CN1, 679–683) is a commentary on religious beliefs and humanity's creation of its own gods; the narrator cites the remark of "un sceptique de génie" who stated, "Dieu a fait l'homme à son image, mais l'homme le lui a bien rendu" (CN1, 679).[1] After detailing the battle between Saint Michael and Satan as the Norman folklore preserved the tale, the narrator concludes with the sardonic remark that "Un autre peuple avait rêvé autrement cette bataille." (CN1, 683), giving the reader to understand that veracity in the spiritual is a matter of regional definition.

All of "Conflits pour rire" (CN1, 426–430) has to do with the comical treatment of quarrels between the Church and those representing secular interests. One of the debates concerns the teaching of Bible stories in public schools; the text seems to take sides in citing the arguments of the secular camp:

> Mais si l'instituteur est autorisé à exposer les aventures de ce recueil d'anecdotes merveilleuses qu'on appelle l'Ancien Testament, peut-on exiger qu'il donne comme articles de foi la création du monde en six jours...?
>
> Quand il apprendra aux futurs élécteurs à ne pas croire aux baguettes de coudrier des sorciers, leur racontera-t-il le miracle à la Rambuteau de Moïse produisant de l'eau par un moyen qui, aux termes de la Bible, ne semble guère anormal? (CN1, 429)

The text goes on to discuss the status of Bible stories as mythology, wondering for pedagogical purposes how to distinguish between mythologies of Christian and pagan origin.

But it is not all to the favor of those who oppose the Church's philosophy. In "Mon oncle Sosthène" (CN1, 503–509) the *libre penseur* narrator plays a practical joke on his freemason uncle which ends up backfiring on him. The nephew sends word to a Jesuit that the freemason is ill and requests a priest; the uncle receives the priest's visit and befriends the priest; when he converts to Catholicism, he leaves all his money to the Church and disinherits the nephew.

In view of all the textual evidence taken together, we must conclude nonetheless that the overall position of Maupassant's novelistic world view is one of materialism over spiritualism, of empiricism over religious faith, of reason over any other manner of revelation. For lack of verifying evidence, the texts in the aggregate take the position that existence is a material phenomenon and that the understanding or interpretation of life is relevant only when limited to material terms. Narrators and characters who speak for the texts' points of view do not tend to be hostile towards Church and clergy, but where science or philosophy is in the balance, ideological materialism takes precedence over faith or any metaphysical concern.

Chance

In chapters 4 and 5 we had occasion to discuss the issue of random chance and its role in Darwinian theory. It figures as a motif in a text discussed just above, "Magnétisme" (CN1, 406–410), in the way through which the secondary narrator debunks one of the purported cases of telepathy. The straw man he sets up is the tale of a young child's waking in the middle of the night, crying that his fisherman father had been killed at sea; in fact, the next day it is learned that the father had been killed. On the face of it, the child had certainly had a telepathic experience, but upon investigation, the storyteller discovered that throughout the entire village, some child would awaken with the same experience at the rate of once a week. So the logical explanation occurs to him who would see it:

> Or, si une de ces fréquentes prédictions coïncide, par un hasard très simple, avec une mort, on crie aussitôt au miracle, car on oublie soudain tous les autres songes, tous les autres présages, toutes les autres prophéties de malheur, demeurés sans confirmation. (CN1, 408)

The coincidence of "telepathy" and actual event is one of chance occurrence, and one would have to say, from a statistical perspective, that it would happen eventually, given the rate of children's nightmares and the danger of their fathers' occupation. By the same token, this is a perfect illustration of the juncture of the empirical mindset with the factor of randomness in the universe, hence the attention paid to the question of chance in a Darwinian reading such as this. The principle of random event governs the process of genetic mutation which Darwin placed at the center of his model of natural selection; without random event, the concept of evolution is impossible.

Other events of chance have come up in the preceding discussion of Maupassant's thematics. We have seen in "Une page d'histoire inédite" the case of the young Napoléon Bonaparte who is saved from an ambush by the chance gesture of a woman to detain her husband for a moment ; this was phrased by the text's narrator as "le grand hasard des événements qui gouvernent les hommes et le monde" (CN1, 186).

There was also the awareness on the part of the negligent father in "Un fils" that his shiftless son could well have developed into something different for the better if chance had given him different circumstances. What the father ruefully refers to as the "terribles lois de l'hérédité" link him to his son:

> ...il est moi par mille choses, par son sang et par sa chair, et ...il a jusqu'aux mêmes germes de maladies, aux mêmes ferments de passions. (CN1, 425)

In his own case, the father knows that heredity has fostered success (he is a member of the Académie française), but for his son it has not worked out so well. In the final analysis, heredity is one force which "predestines" offspring to some extent, but its power may be offset by random chance which can change the circumstances of life at any level (genetics, environment, etc.).

Finally, the motif of chance is related at one point in the corpus to the origins of life, and this in a way that is most germane to a Darwinian reading. The entire narrative of "L'Homme de Mars" (CN2, 1003–1010) has as its premise the assumption that life could originate and evolve by chance elsewhere in the universe, and by the same probability through which it evolved here on earth.

Population concentration vs. isolation

The matter of population concentration and the isolation of populations on the periphery of areas of concentration is another point which Darwin raised in the discussion of speciation (the formation of new species from old). To relate this to the metaphor of the "tree" of life, it is the point along a branch where a twig separates from its ancestral stock and branches out on its own to start a new species distinct from the parent species. Darwin's theory simply put is that population sub-groups, isolated from the geographic concentration of the species, tend to evolve separately since they are free to adapt (genetically and/or behaviorally) to new conditions without the leveling influence of a large, established genetic pool. Separate adaptation is the first step of evolutionary change. To this topic of geographic distribution Darwin devoted chapters 11 and 12 of *The Origin of Species.*

One recurring setting which this brings to mind in the fiction of Maupassant is Corsica, particularly in the way that it is described whenever it comes into play. We see it in the honeymoon sequence in *Une vie* and in numerous short stories, depicted as a wild land, severed geographically from the continent and hence impervious to civilizing influence. Where human experience is concerned, the rules of life are generally not those of France or Italy; the island has evolved its own code for conduct, adapted to the wild terrain and its isolation from civilization. The wilds of Corsica provide an environment for savage behavior which redefines the struggle for survival. Whenever Corsica is the setting for action, it is described as the hiding place of renegades and fugitives, a place where life, if untouched by culture, is at least free and isolated.

In "Le Bonheur" (CN1, 1239–1245) Corsica is the refuge of the daughter of a wealthy French family who runs off with her lover to live happily in poverty and crude conditions. On the one hand, environmental change necessitates for her an adaptation in behavior, as now she must toil to survive. On the other hand, geographic isolation permits her to live an existence free of the social corruption which she is fleeing in favor of a life of happiness.

Similarly, in "Un bandit corse" (CN1, 436–439) the wilds of the island set the scene for adaptation of behavior to suit the conditions. In this short story a timid boy grows to become a fierce bandit after he carries out a vendetta—a most vivid illustration of the principle of adaptation to the environment. "Une vendetta" (CN1, 1030–1034) is yet another case of adapted behavior; in this text, as we saw above, a widow trains her dog to attack on command so that she may avenge the death of her son. Here woman and dog together effect a behavioral change that is in keeping with the unruliness of the surroundings, which latter circumstance the widow exploits for her own purposes, in a way that would not be possible in the novelistic world that Maupassant creates in settings on the mainland.

The description of the setting and its population of fugitives in "Histoire corse" is nothing short of remarkable (CN1, 319–321). These pages detail the narrator's trip on horseback from Ajaccio to Bastia, through the Niolo valley ("qu'on appelle là-bas la citadelle de la liberté" CN1, 319) which underscores the political—and, by extension, the evolutionary—isolation which the countryside affords. The descriptive passage culminates in the depiction of "une fantastique forêt de granit rose" whose boulders present to the eye a stunning assortment of illusions. The fact that these figures have been "rongées par le temps, par la pluie, par les vents, par l'écume salée de la mer", coupled with the fact that there is in the narrator's opinion "rien de plus curieusement ouvragé par le hasard", links together the Darwinian motifs of chance with the cumulative effect of nature's slow forces. With this setting established, the narrator goes on to recount the tale of Calabretti the bandit, followed by the anecdote of the young woman who (as in a similar episode in *Une vie*) requests that she be sent a revolver with which to kill her offensive brother-in-law. Such folklore and behavior, strange to the continental narrator, take shape in a geographic isolation that permits of the evolution of a code of conduct quite different from that of neighboring lands, but which is by the same token suited to circumstances on its island.

In the discussion of speciation, it occurs to one doing a Darwinian reading of Maupassant that there is a subset of the fictional corpus which raises interesting questions about population and isolation. While some of the more than 300 short stories are set in the cities of Paris or Rouen, and some are set in the Franco-Prussian war, and an exceptional few are set in southern France or Corsica or Africa, a great many are set in the Norman countryside. Maupassant makes a great thematic matter of describing Norman peasant society in its eccentricities, and with the volume of texts devoted to it, the characterization of the setting in Normandy accumulates considerably through the corpus. What can be made of this from a

Darwinian standpoint?

The topic of speciation in its finer points (the emergence of new species) was one that perplexed Darwin himself and continues to spark discussion among evolutionary biologists. Everyone knows what a species is (or at least, we think we do, until the devil appears in the details); most frequently the accepted criterion is ability to interbreed. Furthermore, everyone follows the metaphor of the "tree" or bush of life, in which species are envisioned as twigs that branch off unidirectionally from parent stock; but metaphors only go so far, and do not correspond in all ways to the tenors they denote. In actual ranges of habitation and in chronological study, species are harder to pin down than they might seem beforehand. Darwin was of course aware of the difficult subtleties that have vexed taxonomists and evolutionary biologists (and philosophers of science) since the theory of evolution first gained currency. Prior to that point, species were understood to be static in morphology, but with the advent of the notion of change in form over time, the matter of differentiating morphologically and chronologically between related groups became complicated. From chapter 2 of *Origin* onward, the question of differentiating species is at the fore in Darwin's theory.

Darwin and most of the biologists and taxonomists since Darwin have accepted that a new species emerges from a variety of a parent species. The point is important: in a simplified model, a parent species occupies a geographic range in concentrated numbers; eventually, on the periphery of this range, varieties of the same species may appear—different in one or more characteristics, but still recognized as the same species as before; with the passage of enough time and with geographic isolation from the rest of the species, the variety is thought to develop on its own to the point where it no longer has reproductive contact or compatibility with the original species. At this point, all would agree that speciation has occurred. But it is the continuum of change that raises the interesting questions. At what point along the line does change become taxonomically significant? When precisely is a new species formed? How can we precisely distinguish between variety and species?

Fortunately, we need not here try to resolve tricky issues like these. For the purposes of this study, it is instead profitable to dwell on the notion of the "variety" in population as we take up the characterization of Norman peasant society in Maupassant. Motifs borrowed from the Darwinian model, such as isolation and periphery, and the development of behavioral adaptation in a sub-group, seem interesting at this juncture. But let us be clear as to our interest. First, in no way do we imply that speciation itself is in question; instead, we are looking at the effects of geographic and social isolation in the development of a code of conduct (even of morality, for our literary purposes) on the periphery of textual French society. Second, we recognize that we have already found the model of "Social Darwinism" to be pertinent and useful in texts set in urban surroundings where a complicated socioeconomic infrastructure effectively takes the place of a natural environment for a Darwinian model; here, however, we appeal to a more purely Darwinian motif as our interpretive model, viewing the Norman peasant society as a variety of the human species (not, we emphasize, biologically speaking) for the eccentricity and peculiarity that define their portrayal in Maupassant. It is more the strictly Darwinian notion of population dynamics than a Social Darwinist (i.e. economic), transposed model that is now in question.

This is the tenet of evolutionary theory which Darwin named the "principle of diversification." According to this principle, successful species adapt to diversified places in the economy of nature, diverging from similar life forms to create a branching pattern of relationship (Ridley, 14). In the context of natural selection, diversification occurs in response to pressures of the environment: note that Maupassant's Norman stories, to a one, take as a background of economic hardship to which peasant characters respond with behavior and adapted values born of their perception of precarious economic security. In addition to this, the geographic setting of Normandy is unfailingly presented as geographically isolated from the civilizing influence of urban agglomeration.

One thing that invites consideration of the peripheral social setting is the linguistic depiction of Norman society. Maupassant often has recourse to a transcription of Norman dialect, which of course has the effect of differentiating Normans from speakers of the standard French which constitutes the narrators' discourse. One example from among the innumerable will make the point. In "Une vente" (CN1, 1207–1212), a farcical tale in which one drunken Norman bumpkin agrees to the sale of his wife to his drunken friend. To effect the sale by weight, they endeavor to measure the woman's mass by water displacement; subsequently she hauls the two of them into court for depositing her in a barrel of water. Some of her remarks in court are transcribed in the narrative as follows:

> Alors i m'dit: "Ouvre l'oeil et fais comme mé", et le v'la qui s'en va quérir l' grand baril défoncé qu'est sous la gouttière du coin; et pi qu'il l'apporte dans ma cuisine, et pi qu'il le plante droit au milieu, et pi qu'il me dit: "Va quérir d'l'iau jusqu'à tant qu'il sera plein." (CN1, 1208)

The main effect is comical here, but there is also the effect of emphasizing the isolation of the social sub-group. Then too we must consider the appearance of transcribed dialect in other, decidedly serious narratives (for example, "Le Port," CN2, 1125–1133, where the conversation between brother and sister in a brothel is rendered in Norman patois). Humor is obviously not the only effect to be had in the narrative use of this device. It calls attention to the fact that this segment of society is distinct from the rest, in language as in much of the rest of life.

We have already had occasion to concentrate on the narrative juxtaposition of species in "La Ficelle" (chapter 3 above), in the context of discussing the relatedness of species among the premises of Darwinian thought in general. But there is another facet to the motif of mingling human and beast as we saw in the example of "La Ficelle." Depicting the peasants of Normandy in habitual proximity to their livestock has the added effect of implicitly distinguishing them from city dwellers who "herd" together, and certainly from the narrator and implied listener for whom textually identifying with cows and chickens is clearly narrated as a strange and somewhat unpleasant phenomenon:

> Sur la place de Goderville, c'était une foule, une cohue d'humains et de bêtes mélangés. Les cornes des boeufs, les hauts chapeaux à longs poils des paysans riches et les coiffes des paysannes émergeaient à la surface de l'assemblée. Et les voix criardes, aigües, glapissantes formaient une clameur continue et sauvage que dominait parfois un grand éclat poussé par la robuste poitrine d'un campagnard en gaieté, ou le long meuglement d'une vache attachée au mur d'une maison. (CN1, 1081)

As if the description itself were not enough to communicate a narratorial opinion, the very next words in the text are:

> Tout cela sentait l'étable, le lait et le fumier, le foin et la sueur, dégageait cette saveur aigre, affreuse, humaine et bestiale, particulière aux gens des champs. (CN1, 1081)

Adjectives like "bitter" and "awful" make the point for narratorial perspective, but the phrase "peculiar to people of the fields" is more what we are interested in here. The rustic subset of society, with its distinctive traits, stands apart from the rest of the human species, not only geographically in their province, but in appearance, in mores, in tastes, and so on. Even their bodies are physically adapted by acquired traits peculiar to their collective occupation:

> Les mâles allaient, à pas tranquilles, tout le corps en avant à chaque mouvement de leurs longues jambes torses, déformées par les rudes travaux, par la pésée sur la charrue qui fait en même temps monter l'épaule gauche et dévier la taille, par le fauchage des blés qui fait écarter les genoux pour prendre un aplomb solide, par toutes les besognes lentes et pénibles de la campagne. (CN1, 1080)

Everything about them that catches the narrator's eye sets them apart from the rest of humanity in his view. Ultimately, this sequestering permits the evolution of a code of values and mores peculiar to them, and there is very little indeed therein of which the narrators of the Norman tales appear to approve, very little indeed which those narrators appear to consider as normal by standards of the rest of human society. This is what a Darwinian reading finds to be a delineation of a variety of species, emerging from isolation on the periphery of the range where the rest of the species lives in concentration.

A few examples, from among very many possible, of the narrated eccentricity of the isolated Norman sub-population and the attendant, implicit narratorial disdain for the inhumanness of its evolved code of behavior, will suffice to illustrate the point. It is a fair generalization to say that each of the short stories set in rural Normandy highlights a set of values or a character trait that passes for aberrant in the implicit judgment of narrator and implied reader. The narratorial process that communicates that judgment is a skillful and subtly ironic one, for the narrator never utters an explicit condemnation of behavior or values. Yet there is no doubt of what the text holds as an accepted code of values, so it would seem clear that there is a tacit bond between narrator and narratee assuming that storyteller and listener share values up to which those of the Norman peasants are held for comparison. Time after time, that comparison shows the Norman peasantry in an unfavorable light and points up the great disparity between values portrayed in the action and values assumed held by the implied reader. With his tone, the narrator puts such a distance between himself and his characters that the gap between the characters and everyone else (narrator, narratee and the rest of humanity, it seems) appears to be a tribal divide, a separation by ethnic barrier that signals the equivalent of a Darwinian population variety where mores are concerned.

In a well-known short story, "Le Vieux" (CN1, 1130–1137) a farming couple is making its plans for the harvest while the woman's father is on his deathbed about to breathe his last. The conflict for them is the timing of events, the harvest and the coming funeral with its dinner for guests. It is agreed by all that the old man's death cannot be long in coming, so plans are made for funeral and dinner, and invitations are made. Guests arrive expecting a funeral, but the old man has not yet died. The hours pass and still the man lingers, so it is decided to serve the dinner.

> Maître Chicot, la bouche pleine, prononça:
> "S'i nous véyait, l'pé, ça lui f'rait deuil. C'est li qui les aimait d'son vivant."
> Un gros paysan jovial déclara:
> "I n'en mangera pu, à c't'heure. Chacun son tour."
> Cette réflexion, loin d'attrister les invités, sembla les réjouir. C'était leur tour, à eux, de manger les boules.
> Mme Chicot, désolée de la dépense, allait sans cesse au cellier chercher du cidre. (CN1, 1136)

During the dinner, someone notices that Mme Chicot's father has finally died, and with no show of sadness, the Chicots make plans for another dinner after the funeral on the following Monday, regretting only the added expense. The rising spirits of the company, the unattended death of the old man, the calculating nature and the miserliness of the Chicots, all combine to present a scene which the reader is sure to find distasteful without any explicit judgment on the narrator's part.

"Le Petit Fût" (CN2, 77–82) has already figured in our discussion. Here, a calculating peasant (again, named Chicot) is eyeing up an old neighbor woman's land, and he gets her to agree to sell it to him with the proviso that she be allowed to live out her life on the land and collect a monthly payment from him. Tiring of the expense and the wait, Chicot undertakes to hasten her death by leading her to become dependent on the fine alcohol, "la surfine," that he is always on hand to offer her. Her alcoholism causes her death, and Chicot's only reaction (upon which the text closes) is:

> "C'te manante, si alle s'était point boissonnée, alle en avait bien pour dix

ans de plus." (CN2, 82)

The reader hears the irony in that cynical remark, for Chicot is also saying that if he hadn't precipitated her death, the old woman would have been collecting on their bargain for another ten years. Chicot, little more than a murderer whose motive is land and whose poison is good spirits, shows no remorse, and the entire narrative passes for a regional peculiarity.

Maupassant's perception of Norman avarice is the central motif of "L'Aveu" (CN2, 192–197). A single country girl admits to her mother that she is pregnant from exchanging sex for free wagon rides to take the family's farm products to market. After the expected upbraiding, the mother inquires as to the financial savings realized, and ends up advising the daughter to keep the pregnancy secret in order to continue the profitable arrangement for free rides. As usual in these circumstances, the narrator recounts the action in which finances trump morality and conscience, and the whole appears as an aberration of human conduct peculiar to the Norman farmlands. Without saying so explicitly, the narrator leads the listener to see the depicted peasant mores as inhuman and bizarre, certainly not befitting civilized people. The gap between the civilized "us" and the preposterous "them" is the sign of a provincial isolation that is more than geographic alone.

We return to the motif of death hastened for profit's sake in "Le Diable" (CN2, 769–776). The narrator even includes the phrase "avarice normande" (CN2, 769) by way of motivating action. A farmer, whose mother is dying, hires a nurse to watch her while he harvests his crops. He strikes a bargain to pay the nurse by the job instead of by the day, but the nurse has no intention of being taken for less than her time is worth to her. After three days of waiting for the old woman to die, the nurse disguises herself as the devil and frightens her charge to death. In the end, the son regrets the difference between what he did pay and what he would have paid per diem, and the nurse is seen calmly praying for the deceased. In this text the customary motifs of avarice and lack of conscience are joined by a motif of bizarre superstition, combining to form a picture of Norman rustic life as quite otherworldly. For our purposes in a Darwinian reading, the indicated interpretation of the characterization in the Norman short stories is of a breed apart, not of another species, but of a variety of humanity cut off from the rest by great difference of mores, as well as by geographic isolation.

In summary, then, we deduce a consciousness in Maupassant of the implications of population grouping and isolation. In general in the corpus, where populations or small groups are isolated on the periphery of civilization or urban agglomeration, the evolution of new codes of conduct tends to occur. In the texts set in Corsica and North Africa, this is clearly the case; where the great number of stories set in Normandy is concerned, it is possible to trace in detail the appearance of a divergent code of behavior. The consideration of population dynamics dovetails with the two remaining topics for this chapter, the question of race as it appears in the texts, and the interplay of environment and behavior there.

Race

The biological question of race has, as of the late twentieth century, been made quite moot with the advent of DNA typing. In short, DNA study has demonstrated that there is no appreciable genetic difference between human groups traditionally considered separated by geographic, ethnic or visible divisions; it has been shown, for example, that there is more genetic divergence between eastern and western Africans than there is between eastern Africans and *anyone else on earth.* Biology as a science has adopted the position that traditionally considered differences of "race" are without genetic foundation.

Darwin was, of course, without this information, as was Maupassant. In *The Descent of Man,* Darwin devotes considerable space to the treatment of race in the human species; its seventh chapter is entitled "On the races of man," and before deferring a final verdict on the matter of the development of the "races," Darwin attempts an open-minded study of physiological and anthropological characteristics of groups which, on some level, he posits as distinct to some degree. Darwin finds that the study of similarities and perceived differences between human subgroups is inconclusive, and (undoubtedly in response to expectations born of received prejudice universally accepted as truth in his society) he demurs in decisively declaring racial differences as varieties instead of separate species, and claims the issue cannot yet be resolved. In this area, for some reason (probably, again, in response to the prejudice of his day), he does not accept interbreeding as a definitive standard for speciation. But after all his hedging, Darwin concludes in his *Descent* that perceived physical differences between human groups are probably best ascribed to sexual selection, and that there is no indication of separate origins for human subgroups. Furthermore, without explicitly saying so, he gives the reader to assume that he considers the subgroups as varieties and not as separate species. As to his opinion of society's handling of perceived differences dubbed "racial," he elsewhere states his abhorrence of the institution of slavery (Ridley, 6–7, citing Darwin's autobiography).

Any Darwinian reading of Maupassant would, in light of this and of the attention which anthropological difference attracted in late nineteenth-century French society, expect to find evidence in Maupassant's fiction of the motif of race, that is, if a Darwinian reading of Maupassant is at all warranted. Such evidence is there, and it pertains to questions both "racial" and ethnic, To state the issue starkly, if a novelist were to subscribe to the brand of Darwinism that fed into Spencerian Social Darwinism and later into the eugenicist movement in Europe and the United States, one would sow a racist motif into his fiction; if, on the other hand, he were to follow Darwin's own lead and incline towards the acceptance of common biological origins for all human subgroups, his fiction would sooner display an acceptance of what is presently known as diversity. Maupassant's tendency is the latter.

This is most visible in the short story "Boitelle" (CN2, 1086–1094), in which Boitelle, a French soldier stationed in Le Havre, becomes infatuated with a young African woman employed there as a café waitress. Hopelessly in love, he courts her and proposes marriage contingent on his parents' approval. She accepts and agrees to accompany him to the Norman countryside to meet his parents. The parents find her to be a fine young woman and say to their son that they would certainly approve of the marriage except for the fact that she is black-skinned. In the end, the parents withhold their assent, Boitelle breaks off the engagement, and spends the rest of his life regretting the loss. The motif and implicit stance of the text could not be more clear, for the narrative gives an entirely favorable characterization of Boitelle, and the action openly shows the working out of a racist bias and the human cost it takes.

"Allouma" (CN2, 1095–1117), which we have already discussed above, and "Mohammed-Fripouille" (CN2, 333–340) both bring into play the motif of colonists' attitude toward indigenous Arab peoples. In both cases, the attitude of unquestioned and inherent superiority of Europeans stands unchallenged, but less as a function of racial divide than of political order and social custom. In "Allouma" the narrative makes use of a framing device to put some moral distance between a primary narrator (who appears as a young traveler, unjaded by the colonial disdain for the native people) and a secondary narrator, Auballe, who unabashedly recounts his own tale of reprehensible and ungentlemanly appropriation of an Arab woman whom he treats as something of a slave and concubine. Given his caddish behavior and his outright pride in that conduct as he narrates his story to the primary narrator, Auballe's high-handed chauvinism does not stand forth as a positive code of values offered by the text as laudable or heroic in any way. And given Maupassant's general tendency to keep narrator and scriptor mum in the face of values of which they clearly disapprove, and in lieu of explicit judgment (preferring the devices of understatement and irony), it is a fair argument to make that Auballe's racism stands right alongside his sexism as part of a general disposition which is crafted to appear distasteful to the reader; with that, enough is said for Maupassant's style to relegate Auballe's values to a category of curiosities not suited for more civilized tastes. Emblematic of Auballe's beliefs is his remark that he had become attached to "cette créature d'une autre race, qui me semblait presque d'une autre espèce, née sur une planète voisine", and that Arab women "...sont trop près de l'animalité humaine" (CN2, 1107). His listener says nothing in reaction throughout Auballe's long story, and that story is permitted to stand as a tale of "un amour dans le désert" (CN2, 1117), but told by a boor who goes out of his way to show himself as such. So while the accepted attitude of colonial racism is not overtly repudiated, it is by no means held forth as a laudable code of values or set before the reader as a commendable posture.

"Mohammed-Fripouille" is more ambiguous on this issue. The narrative of a *légionnaire* captain Marret is again framed within the first-person discourse of a primary narrator, and the storytelling again goes on in the setting of a Frenchman's home in Algeria, as is the case in "Allouma." But in this case, the primary narrator (as listener for the inserted narrative) says very little to indicate his alienation from his host's hardened attitude toward the dominated native people. When the guest asks to hear the captain tell a story from among his experiences, Marret tells of Mohammed-Fripouille, a Turkish junior officer in the Legion, who volunteers to avenge the death of an English traveler at the hands of an Arab tribe. With six other legionnaires, Mohammed-Fripouille manages to surprise and capture fifty tribesmen whom he has bound around the neck in a line which he calls "la chaîne arabe" (CN2, 339). After the legionnaires have eaten and as they are about to lead the prisoners off, a group of Arab women attack. In what Mohammed-Fripouille calls "le dessert" (CN2, 340), he orders the troop to saber them. At the end of the inserted narrative, the primary narrator takes over the discourse only to say that

> Je ne répondis rien. Je songeais à l'étrange pays où l'on pouvait voir de pareilles choses; et je regardais dans le ciel noir le troupeau innombrable et luisant des étoiles. (CN2, 340)

This listener's reaction to a tale of fierce vengeance taken on an entire group is one of mild surprise at the strange goings-on in a country strange to him, and here there is nothing implicit to indicate a thought of moral censure on his part. By the same token, there is nothing to indicate approval, either, so the racist motif is allowed to stand without comment in this case.

Of course, there is in French history an undercurrent of another variety of ethnically defined prejudice, one that would explode, shortly after Maupassant's death, into a political and social scandal that would see the leader of the Naturalist school, Emile Zola, take a courageous public stand on the side of justice. Anti-Semitism had roots in western Europe that extended to the Middle Ages, and it was to linger well beyond the Dreyfus affair. If an author of any stripe were to sow a thread of eugenics-based racism into a corpus of literature, this would be the likely matter (consider, for example, the writings of the infamous Gyp, *nom de plume* of Sybille-Gabrielle Marie-Antoinette de Riquetti de Mirabeau whose crude nationalism and anti-Semitism made her career). Maupassant, for his part, would have nothing to do with this vein of prejudice in the thematics of his fiction.

Readers familiar with *Bel-Ami* alone might not be persuaded; it is true that the figure of M. Walter is presented as rather one-dimensional in his seriousness as a businessman, but there is no allusion in the narrative to ethnically-based prejudice on the part of characters or narrator; Walter's vanity as an art collector is not flatteringly portrayed (R, 294–295), but the rather unscrupulous protagonist of the novel notes only in passing and without apparent prejudice that Mme Walter, whom he intends to seduce, is not herself Jewish (R, 381). The question of ethnic background in *Bel-Ami* is rather inert.[2]

It is in *Mont-Oriol*, however, where the motif of anti-Semitism receives overt treatment. Venture capitalist William Andermatt, husband of the protagonist Christiane, is an ambitious and energetic businessman who happens to be Jewish. Gontran, the freeloading brother of Christiane, takes frequent pot shots at Andermatt in conversation throughout the novel, from behind the mask of apparent bonhomie. Finally, when Andermatt has had enough of Gontran's ironic remarks, he confronts him with an eloquent self-defense and condemnation of anti-Semitism (R, 631–632).

Elsewhere, Jewish ethnicity is a matter of exotic attractiveness. In "L'Inconnue" (CN2, 442–447) the framed narrative—told in response to another Gontran, as it would happen—is the story of an infatuation with a woman seen passing by on the sidewalk. Roger des Annettes tells of seeing a striking woman several times and finally approaching her. In view of her exotic charm, Roger assumes that she is Jewish. He says of her,

> Il me sembla que je voyais une des magiciennes des *Mille et une Nuits*, un de ces êtres dangereux et perfides qui ont pour mission d'entraîner les hommes en des abîmes inconnus. ...
> Qui est–elle! Une Asiatique, peut-être? Sans doute une juive d'Orient? Oui, une juive! J'ai dans l'idée que c'est une juive. Mais pourquoi? Voilà! Pourquoi? Je ne sais pas! (CN2, 446–447)

For this character, Jewish ancestry is anything but grounds for prejudicial discrimination, but instead is a most appealing trait. In the end, it appears to be Maupassant's inclination not to partake of the fashionable anti-Semitism of his day, and instead to downplay the "racial" differences between sub-groups of human society. This is, coincidentally or not, in keeping with a Darwinian view of nature.

At this juncture we should make a passing reference to the avenue of thought known as "eugenics," often associated with the topics of race and ethnicity. The term "eugenics" was originated by Darwin's cousin, Francis Galton, who took it from a Greek root for "good in birth" (Caudill, 98–99). Ostensibly, the movement, which generated much interest among the upper socioeconomic castes in Europe and the United States, was to encourage "desirable" types to procreate and discourage (or disallow) the procreation of "undesirable" types—desirability being defined along lines of race, family background and measures of intelligence, for the most part. Edward Caudill (96–113) has written an excellent summary and evaluation of this misguided abuse of pseudo-science, under the rubric of "Misuses of Darwinism." At any rate, it was mentioned above in chapter 5 that Maupassant had nothing to do with the movement or its beliefs, and that the closest he came to its thematics was in the short story "Berthe" (CN2, 355–364) in which a young woman of undeveloped intelligence is married off by her father in the hope of directing positively her appetites and passions into motherhood; the idea ends in disaster. Given the implications of the text's action, and the fact that it is narrated by the family's physician, it may be construed as a cautionary tale about encouraging marriage or reproduction among the mentally handicapped. This is still a long way from the racial and social policy aspirations of the eugenicists.

The only other Maupassant text which deals even marginally with thematics related to eugenicist thought is "L'Endormeuse" (CN2, 1159–1168), a rather light, "modest proposal" for an approved institution for assisted suicide. This is the only instance of the motif of euthanasia in Maupassant.

Environment and behavior

The final topic for our consideration in this reading of Maupassant's fiction through the lens of evolutionary theory is that of adaptation in behavior brought on by environmental pressure or change. One obvious subset of the corpus in which this matter calls out for attention is the group of short stories set in the context of the Franco-Prussian War. In a word, the setting of war brings out the very worst in characters for whom the struggle to survive is rendered critical by the atmosphere of mayhem which either battle or occupation forces upon them. In some of Maupassant's more widely read texts, the war elicits acts which clearly would not occur otherwise: in "Mademoiselle Fifi" it provokes murder; in "Boule de Suif" it leads the prim bourgeoisie to coerce the title character to compromise herself.

Change in the conditions of the environment can either be to the detriment of the individual (or to the species) or they can prompt adaptation which enhances chances for survival. This is classical Darwinism. In "Première neige" (CN1, 1094–1102) a vital Parisian woman marries a benighted Norman who will not consider moving from his Norman castle. With the onset of winter, the woman suffers from both the cold and the boredom that she has never before been called upon to endure. Even after going south temporarily to recuperate, she knows that she will die upon returning to the harsh Norman climate. Narratorial tone is the only textual indicator we have to arrive at an implied judgment: the narrator seems to pity the woman and to hold her dullard husband in scorn.

Harshness of the physical environment in which characters find their lives changed is, of course, a catalyst for drastic action. "L'Horrible" (CN2, 114–119) is the inserted narrative recounting an episode in the North African desert of the Sudan: with the ambush of the greater part of an exploratory mission and the loss of animals and supplies, the survivors are pushed to the limit of endurance; finally one of them kills another, and the rest of the party resorts to cannibalism in order to survive. But such peripeteia of climate and terrain need not be the case for drastic effects in the lives of characters, as in other texts there are social and economic changes that have severe consequences.

"La Parure" (CN1, 1198–1206) is a well-known example of economic loss that necessarily brings behavioral adaptation in its wake. In order to survive the loss of the borrowed necklace (and the financial hardship imposed by the task of replacing a diamond necklace), Mathilde must alter her lifestyle completely, submitting to manual labor, domestic chores and poorer living quarters. The change has the effect of toughening her, certainly, but it also appears to age her visibly. In the end, she survives, but not in a way to her liking. Any pretensions of aesthetic quality are gone from her life which becomes one of physical survival only. Environment here is the Social Darwinian one of economic reality in the urban context. The study of Sharon P. Johnson looks further into the question of contamination of the environment in Maupassant and the interplay between nature and the "ills of the metropolis", relating the topic metonymically to sexuality and social mores (165).

The pressures exercised by this sort of setting drive characters to desperate measures. In several texts (e.g., "Les Soeurs Rondoli," CN2, 133–161, and "L'Odysée d'une fille," CN1, 197–1003), women are driven to prostitution as a last resort in order to stay alive. This is a response to the broad environment depicted in economic terms. Other texts create closed environments whose pressures are either reduced microcosmically (e.g., a working environment where the need to retain employment constitutes a central conflict that speaks for the economic realities in the world outside) or emotional in nature.

An example of the first sort of circumscribed environment is in "La Question du latin" (CN2, 797–805) where a meddling student takes it upon himself to arrange a rendezvous between his Latin teacher and a young woman, knowing well that such a pastime could compromise the teacher's standing in the school. The contrast between the rarified atmosphere within the school's walls and the much more tolerant ambiance in the world outside those walls serves to underscore the pressure that is brought to bear on the Latin teacher. In the end, he does lose his position to the intolerance of the school environment, but is happy to set up shop as a grocer with the same young woman as his wife.

"Mademoiselle Perle" (CN2, 669–684) is an example of the other sort of self-enclosed environment whose pressures are defined in the interpersonal relationships it contains. In this case, the household of the Chantal couple, their two daughters, and an adoptive sister of M. Chantal live within their house in a self-imposed retreat from the world, in the city of Paris. The narrator, who is invited as an old friend to spend the feast of the Kings with the family, discovers that M. Chantal has hidden since youth his love for Mlle Perle, whom his family adopted as an infant. The narrator, artfully presented as a meddling and inconsiderate intruder, takes it upon himself to disrupt the equilibrium of the family's life by revealing M. Chantal's secret to Perle; in the ensuing pandemonium—Perle faints and Chantal is still emotionally overwrought—the narrator beats a hasty exit. It is clear to the reader that the peaceful life of the Chantal household is sent into turmoil from which it will not recover. The narrator has changed the balance of the environment with his thoughtless experimentation, and has made of the household a sort of laboratory where he undertakes to manipulate people as though they were subjects in a scientific project.

The discussion of the effects of environment in Darwin's theory inevitably comes back around to the centerpiece of Darwin's thought, the great reality of natural selection. It is the environment that posits the ground rules for the struggle to survive, the limitation of resources, the conditions of climate and terrain. Environment is the arena for the competition that plays out, competition of species against the physical world, competition between species, competition among individuals of the same species for resources and for reproductive success. Maupassant's fiction sets all sorts of struggle into motion, and perhaps, as Darwin suggests in the *Origin*,

> ...the struggle almost invariably will be most severe between the individuals of the same species, for they frequent the same districts, require the same food, and are exposed to the same dangers. (126)

When society is the environment, or when the man-made crucible of war is the locus of action in Maupassant, the human animal is made to appear as more germane to other species that walk the earth than to any God, or to the angels, or to any Romantic hero or noble savage that acted on the pages of the literature of a prior time. Conflict in the novelistic world of Maupassant is not of the grand scale that Classical literature would have, nor is it the soulful crisis of the troubled and melancholy Romantic hero. Instead it is the struggle for life, very much in the Darwinian sense, a struggle that may pit person against person, or individual against society, or a man or woman against the world, where the stakes are nothing more or less than life itself. Losers in the struggle fall by the wayside, and it is a pitiless world that does nothing to rescue them. If the thematics of Maupassant do not extend to matters like "evolution" in the strictest sense, the change of species over time, or the extinction of species, the rest of Darwinist thought is demonstrably present in action and descriptive motif in the novels and short stories. Factors like random chance, materialism, population dynamics, environment and adaptation, the topics of this chapter, confirm the suitability of the model of Darwinian thought for the fiction of Maupassant, the Naturalist, the skeptic, the materialist, the observer of society and of mores.

NOTES

1. Louis Forestier attributes this thought to two possible sources, Voltaire and Schopenhauer (CN, 1498, note 2).
2. Louis Forestier claims that Maupassant never showed great consideration for Jewish men (CN2, 1451, note 1 for p. 444), while finding Jewish women attractive. There is no real direct evidence of this disposition in Maupassant's fiction, and it seems that, at least on the theoretical level, Maupassant tends in his works to repudiate anti-Semitic prejudice.

Conclusion

It would be a fair assessment to say that Darwin's materialism colored by Schopenhauer's pessimism exercised an influence on Maupassant's rather dark world view. In the absence of a religious redemption of life on earth, existence in the novels and short stories appears as limited to the struggle for survival that culminates in the extinction of the individual and eventually of the species. Not a rosy picture, but then neither is the general one sketched by this corpus of fiction.

The question of Maupassant's place in literary history is one that I will leave for others to resolve. But part of an edifying conclusion to this study would consist of highlighting some of the ways in which Maupassant points forward in literature. Of course he is held in high regard by later writers in France and beyond its borders. However, it is interesting to note that he makes two significant thematic announcements in his fiction. First, he foreshadows on one notable occasion the great Proustian motif of memory, and particularly the memory whose accumulation is the only, evolving identity of the novel character. I refer to the striking beginning of the text entitled "Malades et médecins" (CN2, 100–106):

> Singulier mystère que le souvenir! On va devant soi, par les rues, sous le premier soleil de mai, et tout à coup, comme si les portes depuis longtemps fermées s'ouvraient dans la mémoire, des choses oubliées vous reviennent. Elles passent, suivies par d'autres, vous font revivre des heures passées, des heures lointaines.
>
> Pourquoi ces retours brusques vers l'autrefois? Qui sait? Une odeur qui flotte, une sensation si légère qu'on ne l'a point notée, mais qu'un de nos organes reconnaît, ...un rien qui nous effleura en une circonstance ancienne et qu'on retrouve, suffit à nous faire revoir tout à coup un pays, des gens, des événements disparus à notre pensée. (CN2, 100)

Memory, sparked by an impression of the senses, brings the past to the present to remind the character of what he or she has been. This fluidity or accumulation of identity as it evolves has been an aspect of Proust that scholars see as a thread leading to the Existentialists. Proust also remarked that a book is "the product of another *moi*", and in no other previous author could the motif of "l'autre qui est en nous" (as in *Pierre et Jean*, "Le Horla" and numerous places elsewhere) be more in evidence than in Maupassant.

Surely the Existentialists saw in Maupassant's materialism and agnosticism a source of thematic influence. Maupassant's novelistic world is one in which character reacts to physical circumstances in the environment, without the influence or the hope of intervention of another plane of existence. Even in Maupassant's stories of the fantastic, the double is a sort of extension of the self, somehow part of the character and somehow alienated from him. These are notions that would appeal in some measure to the Existentialists of the next century. And as we have noted above (in concert with Daniel C. Dennett), to the extent that one can call Nietzsche the father of Existentialism, one can refer to Darwin—and thus in some measure to the Darwin in Maupassant's world view—as the grandfather of that development in thought and literature. In any event, Maupassant's pessimism and his portrayal of the contingency of life are clear links to the Existentialists. Likewise, the Darwinist current through Maupassant also announces the challenge to what Marx called "teleology" (and what we call "Intelligent Design" today) which the Existentialists will pick up.

Interesting motifs bearing upon the human condition have come to light in the Darwinian reading of Maupassant's fiction. Heredity, race, abortion, selection and environment, and the amorality of the universe, to name a few, all figure significantly. The degree to which each of these motifs is presented in the texts in any way sympathetic to humanity's plight betokens an attitude that is much less harsh than Maupassant's general pessimism would lead one to expect. One might say, for example, that heredity in Maupassant, when compared to the motif in Zola, is less absolute a condemnation: of Julien's two sons in *Une vie*, one turns out to be a decent sort.

There is much in the way of future study suggested by this sort of reading of Maupassant. First, one might take as a point of departure the theoretical work of Joseph Carroll for Darwinian literary study (though less for his polemics against contemporary schools of epistemology and literary ontology than for his properly Darwinian theory). Naturally, there is the possibility of the Darwinian reading of other authors and works: Zola and the other Naturalists would be likely candidates, as might be the realist writers, the Existentialists, texts in which the motifs studied here figure prominently, and so on. Other related motifs for study in Maupassant or other writers might include the notion of resources in the environment, that is, an applied analysis of cost in survival and reproduction. One might undertake the close reading of Maupassant texts through the Darwinian lens. While much has been done in literary study (for example, a computer search of the MLA bibliography for "Darwin" brings up 570 hits!), much remains to be done, especially in the study of French literature.

A fascinating topic of literary research would be the collision in literature between Social Darwinism and ethics, the opposition between an abstract theory and what was made of it in transposition to social thought, on the one hand, and the realization that there is more to humane living than the survival of the fittest, on the other. The study of the portrayal of capitalist economics in literature as Darwinian might as well be interesting: not merely Social Darwinism as conventionally understood tooth-and-claw struggle for survival, but Darwinism as a model for interpreting economics and economic process in literature; this would look beyond social structure to topics like the appearance of market niches filled by opportunists who endeavor to survive, the free market as an environment of competition for limited resources, etc. Such a model might be applied to any literary text in which the motif of economics is in the fore, from *Turcaret* to *César Birotteau* to *L'Argent* to narrative Francophone literature. Again, on the theoretical level, one might study the apparent contradiction in which the Marxist model is in harmony with the Darwinian model where ideological materialism and teleology (or "Design") are concerned, but at odds with it where capitalism and Social Darwinism appear as products of a Darwinian reading of history.

What can the Darwinian reading of fiction contribute to culture? It can begin by raising awareness of one of the truly great ideas of modern thought. It can provide one more tool for the meaningful interpretation of literature, and be a key for reading scientistic literature, Realism and Naturalism produced in a milieu enlightened by evolutionary theory. It can dispel the misunderstanding of evolution caused by erroneous received notions or by political disinformation. It points up the interplay between literary art and science and philosophy, and it enlightens the study of literature which approaches its subject matter from the perspective of the history of ideas.

As the present study has tried to show, a Darwinian reading can shed light on particular motifs in narrative fiction. For a motif like that of herd mentality and social sanction in "La Ficelle," this reading offers substantial explanation in place of any other model's reading of gratuitous nastiness. It offers a logical interpretation of adoption and fostering in "Mouche" that goes beyond a reading of incidental charity. It renders a systematic explanation for behavioral adaptation in "Le Vagabond." In general, it offers a historical and comparative situation of Social Darwinism (and all that is bad about it) in a broader ideological current.

One remaining question to be addressed has to do with whether it can be concluded from this reading that Maupassant shared a Darwinian world view: without being a rhetorical exponent of Darwinian ideology, does Maupassant have what Darwin called at the end of his *Origin of Species* "this view of life"?

Certainly as concerns agnosticism, the amorality of the universe, determinism and ideological materialism (right down to the cycle of organic life, as in "Coco" where the horse's remains fertilize the grass above its grave), Maupassant gives evidence of a Darwinian view of nature and the universe. Relative to the central Darwinian tenet of the relatedness of species, Maupassant shared the Naturalists' view of humanity as an animal species; "Le Docteur Héraclius Gloss," "Amour" and "Une vendetta" show animals approximating the lives of humans in their environments. "Le Horla" and several similar textual presences elsewhere signal sentient, intelligent non-human life in Maupassant's novelistic world, yet even these retain material properties (like drinking liquids, turning book pages, and blocking fields of vision). The point here is that *Homo sapiens* is not alone, atop some putative ladder of life, and not a privileged species. "L'Homme de Mars" toys with the possibility of the evolution of life elsewhere in the universe. All of these notions figure in the Darwinian view of life.

Maupassant's depiction of the environment as hostile to life is very much in keeping with Darwin's thought on the subject. The stories set in the Franco-Prussian war make this clear, and the influence of Herbert Spencer is visible in Maupassant's inclusion of motifs of Social Darwinism in the corpus whenever socioeconomic environment is in play. Motifs of natural disaster (avalanche, harsh climate, and the like), harsh terrain, predators and scarcity of food are all Maupassant motifs that are reckoned with in Darwinian thought. In Maupassant, as in Darwin, natural selection is at work: only the strong survive (physically or economically), a heartless, blind and unrelenting environment weeds out the defenseless, and life is a struggle for survival. For Maupassant as well as Darwin, sexual selection figures largely in success at reproduction. And finally, the heredity of traits serves to bestow an advantage, albeit a slight one, in the fictional struggle to survive in Maupassant: the recurring, perhaps favorite, inherited trait for description is the parsimony of the Norman peasantry.

In light of all that we have seen, we must surely conclude that Maupassant's view of life, nature and the universe was greatly influenced by the wave of evolutionary thought that was sweeping over the intellectual world of his times. His depiction of humanity is a complex one, a characterization that seats reason, altruism and the occasional ideal next to bestial impulse, competition and the inexplicable. Darwin's closing observation in his *Descent of Man* is this:

> We must, however, acknowledge, as it seems to me, that man, with all his noble qualities, with sympathy which feels for the most debased, with benevolence which extends not only to other men but to the humblest living creature, with his godlike intellect which has penetrated into the movements and constitution of the solar system—with all these exalted powers—man still bears in his bodily frame the indelible stamp of his lowly origin.

With this, Maupassant would doubtless concur.

Bibliography

Primary Texts

Maupassant, Guy de. *Contes et nouvelles, 1*. Ed. Louis Forestier. Paris: Gallimard, 1974.
———. *Contes et nouvelles, 2*. Ed. Louis Forestier. Paris: Gallimard, 1979.
———. *Romans*. Ed. Louis Forestier. Paris: Gallimard, 1987.

References

Atwell, John E. *Schopenhauer on the Character of the World: The Metaphysics of Will*. Berkeley, Los Angeles, London: University of California Press, 1995.
Bernardini, Jean-Marc. *Le Darwinisme social en France (1859–1918): Fascination et rejet d'une idéologie*. Paris: CNRS Editions, 1997.
Berta, Michel. "Sainte Elisabeth Rousset dite Boule de Suif." *Excavatio* 12 (1999): 121–130.
Carroll, Joseph. *Evolution and Literary Theory*. Columbia and London: University of Missouri Press, 1995.
Caudill, Edward. Darwinian Myths: The Legends and Misuses of a Theory. Knoxville: University of Tennessee Press, 1997.
Clark, Linda L. *Social Darwinism in France*. University, Alabama: University of Alabama Press, 1984.
Conry, Yvette. *L'Introduction du darwinisme en France au XIXe siècle*. Paris: Vrin, 1974.
Darwin, Charles. *The Origin of Species by Means of Natural Selection, or the Preservation of Favoured Races in the Struggle for Life*. New York: Avenel, 1979.
———. *The Descent of Man And Selection in Relation to Sex*. New York: A.L. Burt, c. 1874.
———. *The Expression of the Emotions in Man and Animals*, 3rd edition. New York and Oxford: Oxford University Press, 1998.
Dawkins, Richard. *The Blind Watchmaker*. New York: Norton, 1986.

———. *Climbing Mount Improbable.* New York: Norton, 1996.

———. *River Out of Eden: A Darwinian View of Life.* New York: Basic Books, 1995.

———. *The Selfish Gene.* New York: Oxford University Press, 1976.

Dennett, Daniel C. *Darwin's Dangerous Idea: Evolution and the Meanings of Life.* New York: Touchstone / Simon and Schuster, 1995.

Dobzhansky, Theodosius. "Nothing in Biology Makes Sense Except in the Light of Evolution." *American Biology Teacher* 35 (1973): 125–129.

Donaldson-Evans, Mary. "Maupassant and the Physician-Narrator." *Maupassant conteur et romancier.* Ed. Christopher Lloyd and Robert Lethbridge, Durham: University of Durham, 1994.

———. "Pricking the Male Ego: Pins and Needles in Flaubert, Maupassant, and Zola." *Nineteenth Century French Studies* 30.3–4 (Spring-Summer 2002): 254–265.

Dumesnil, René. *Guy de Maupassant.* Paris: Tallandier, 1979.

Färnlöf, Hans. "Discours et distance: L'Ironie dans trois contes fantastiques de Maupassant." *Cahiers Naturalistes* 77 (2003): 169–180.

Gould, Stephen Jay. *Ever Since Darwin: Reflections in Natural History.* New York: Norton, 1977.

———. *Hen's Teeth and Horse's Toes.* New York: Norton, 1983.

———. *The Lying Stones of Marrakech.* New York: Harmony, 2000.

———. *The Mismeasure of Man.* New York: Norton, 1981.

———. *The Panda's Thumb.* New York and London: Norton, 1982.

Hadlock, Philip G. "Maleness, Madness, and the Origins of Patriarchy in Maupassant's Qui sait?" *Australian Journal of French Studies* 39.3 (September-December 2002): 339–353.

———. "Telling Madness and Masculinity in Maupassant's 'Le Horla.' "*Esprit Créateur* 43.3 (Fall 2003): 47–56.

Henry, Freeman G. "Anti-Darwinism in France: Science and the Myth of Nation." *Nineteenth-Century French Studies* 27. 3 and 4 (Spring-Summer 1999): 290–304.

Hofstadter, Richard. *Social Darwinism in American Thought.* Boston: Beacon Press, 1955.

Jameson, Fredric. *The Political Unconscious: Narrative as a Socially Symbolic Act.* New York: Cornell University Press, 1981.

Johnson, Sharon P. *Boundaries of Acceptability: Flaubert, Maupassant, Cézanne, Cassatt.* New York: Peter Lang, 2000.

Kennedy, James G. *Herbert Spencer.* Boston: Twayne, 1978.

Lefebvre, Paul. "'Le Horla': Une oeuvre de Guy de Maupassant (1850–1893)." (2004) <http: / / www.denise-pelletier.qc.ca / fiches / pieces / horla.html.>.

Lerner, Michael G. *Maupassant.* New York: Braziller, 1975.

Magee, Bryan. *The Philosophy of Schopenhauer.* Oxford and New York: Clarendon Press and Oxford University Press, 1983.

Mamoon, Sayeeda H. "The Self and the Other / the Other Self: Negotiating Alterity in Maupassant's Fiction." *Excavatio* 18.1–2 (2003): 334–347.

Maupassant, Guy de. *Pierre et Jean*. Ed. Marie-Claire Ropars-Wuilleumier. Paris: Albin Michel, 1984.

———. *Selected Short Stories*. trans. Roger Colet. London and New York: Penguin, 1971.

Mills, Cynthia. *The Theory of Evolution: What It Is, Where It Came From, and Why It Works*. Hoboken: John Wiley, 2004.

Morand, Paul. *Vie de Guy de Maupassant*. Paris: Pygmalion/Gérard Watelet, 1998.

Nitze, William A., and E. Preston Dargan. *A History of French Literature*, 3rd ed. New York: Holt, 1938.

Place-Verghnes, Floriane. "Tabou or Not Tabou, dire ou ne pas dire: Maupassant et l'autocensure." *Neophilologus* 88.3 (July 2004): 367–384.

Ponnau, Gwenhaël. "Du Concept de l'origine à la pensée de la fin: sur quelques avatars fin-de-siècle du darwinisme." *Fin de siècle: Terme-évolution-révolution?*. Ed. Gwenhaël Ponnau. Toulouse: Presse Universitaire du Mirail, 1989.

Ridley, Mark, ed. *The Darwin Reader*. New York and London: Norton, 1987.

Schor, Naomi. *Breaking the Chain: Women, Theory and Realist Fiction*. New York: Columbia University Press, 1985.

Sherard, Robert H. *The Life, Work and Evil Fate of Guy de Maupassant*. London: Laurie, 1926.

Smith, Diane M. "The Evolution of the Working Class Novel in Europe: Darwinian Science and Literary Naturalism." *Excavatio* 8 (1996): 72–85.

Stivale, Charles. "Horny Dudes: Guy de Maupassant and the Masculine Feuille de rose." *Esprit Créateur* 43.3 (Fall 2003): 57–67.

Wallace, A.H. *Guy de Maupassant*. New York: Twayne, 1973.

Wells, Spencer. *The Journey of Man: A Genetic Odyssey*. Princeton: Princeton University Press, 2002.

Whyte, Peter. "Maupassant et le réflexe conditionné." *Maupassant conteur et romancier*. Ed. Christopher Lloyd and Robert Lethbridge. Durham: University of Durham, 1994.

Wiltshire, David. *The Social and Political Thought of Herbert Spencer*. Oxford and New York: Oxford University Press, 1978.

Zola, Emile. *Germinal*. Paris: Fasquelle, 1983.

Works consulted

Bancquart, Marie-Claire. "Maupassant et Paris." *Revue d'Histoire Littéraire de la France*. 94.5 (September-October 1994): 793–799.

———. "Maupassant et l'argent." *Romantisme: Revue du Dix-neuvième Siècle* 13.40 (1983): 129–139.

Berra, Tim M. *Evolution and the Myth of Creationism: A Basic Guide to the Facts in the Evolution Debate*. Stanford, CA: Stanford University Press, 1990.

Brooks, John I. *The Eclectic Legacy: Academic Philosophy and the Human Sciences in Nineteenth-Century France*. Newark, DE: University of Delaware Press, 1998.

Bury, Mariane. "Maupassant pessimiste?" *Romantisme: Revue du Dix-neuvième Siècle*. 18.61 (1988): 75–83.

Caplan, Arthur L., and Bruce Jennings, eds. *Darwin, Marx and Freud: Their Influence on Moral Theory*. New York: Plenum, 1984.

Champagne, Roland. "Revealing the Ideological Mask of Culture; The French New Philosophers, Maupassant, and Literary Criticism." *Language and Style* 18.3 (Summer 1985): 232–240.

Diamond, Jared M. *The Third Chimpanzee: The Evolution and Future of the Human Animal*. New York: HarperCollins, 1992.

Dissanayake, Ellen. "Darwin Meets Literary Theory." *Philosophy and Literature* 20 (April 1996): 229–239.

Donaldson-Evans, Mary. "Doctoring History: Maupassant's 'Un coup d'état'." *Nineteenth Century French Studies* 16.3–4 (Spring-Summer 1988): 351–360.

———. "The Harlot's Apprentice: Maupassant's *Bel-Ami*" *French Review* 60.5 (April 1987): 616–625.

———. "The Last Laugh: Maupassant's 'Les Bijoux' and 'La Parure'." *French Forum* 10.2 (May 1985): 163–174.

———. "Maupassant et le carcan de la nouvelle" *Cahiers Naturalistes* 74 (2000): 75–82.

Duncan, Janice. "Violence in the Short Stories of Guy de Maupassant." Diss., University of Nebraska, 1992.

Eldredge, Niles. *The Triumph of Evolution and the Failure of Creationism*. New York: Owl/Henry Holt, 2000.

Eng, Steve. "De Maupassant's Romantic Beginnings." *Romantist* 2 (1978): 25.

Farley, John. "The Initial Reactions of French Biologists to Darwin's *Origin of Species*." *Journal of the History of Biology* 7.2 (1974): 275–300.

Flórido, Lúcia. "Le Narrateur du 'Horla': Fou, hanté ou artiste?" *Tropos* 21 (Spring 1995): 13–26.

Forestier, Louis. "Guy de Maupassant et le salon de 1886." *Flaubert et Maupassant: Ecrivains normands*. Ed. Joseph-Marc Bailbé and Jean Pierrot. Paris: Presse Universitaire de France, 1981. 111–125.

———. "Maupassant et le peuple." *Peuple, mythe et histoire*. Ed. Simone Bernard-Griffiths and Alain Pessin. Toulouse: Presse Universitaire du Mirail, 1997. 191–204.

Freimanis, Dzintars. "Maupassant as a Romantic." *Romanic Review* 54 (1963): 274–280.

Fusco, Richard. *Maupassant and the American Short Story: The Influence of Form at the Turn of the Century*. University Park: Pennsylvania State University Press, 1994.

Gould, Stephen Jay. *Bully for Brontosaurus: Reflections in Natural History*. New York: Norton, 1991.

———. *Dinosaur in a Haystack: Reflections in Natural History*. New York: Harmony, 1995.

———. *Eight Little Piggies: Reflections in Natural History*. New York: Norton, 1993.

———. *The Flamingo's Smile: Reflections in Natural History*. New York: Norton, 1985.

———. *Full House: The Spread of Excellence from Plato to Darwin*. New York: Harmony, 1996.

———. *I Have Landed: The End of a Beginning in Natural History*. New York: Harmony, 2002.

———. *Ontogeny and Phylogeny*. Cambridge, MA: Belknap Press of Harvard University Press, 1977.

———. *Time's Arrow, Time's Cycle: Myth and Metaphor in the Discovery of Geological Time*. Cambridge,, MA: Harvard University Press, 1987.

———. *Wonderful Life: The Burgess Shale and the Nature of History*. New York: Norton, 1989.

Hadlock, Philip G. "(Per)Versions of Masculinity in Maupassant's 'La Mère aux monstres'." *French Forum* 27.1 (Winter 2002): 59–79.

Hainsworth, G. "Schopenhauer, Flaubert, Maupassant: Conceptual Thought and Artistic Truth." *Currents of Thought in French Literature: Essays in Memory of G.T. Clapton*. Ed. J.C. Ireson. Oxford and New York: Basil Blackwell, Barnes & Noble, 1966. 165–190.

Idt, Geneviève. "Bluettes sartriennes." *Maupassant, miroir de la nouvelle*. Ed. Jacques Lecarmé and Bruno Vercier. Paris: Presse Universitaire de Vincennes, 1988. 243–258.

Joly, Bernard. "La reprise de l'incipit dans les Contes et Nouvelles de Maupassant." *Cahiers Naturalistes* 74 (2000): 83–96.

Jones, Steve. *Darwin's Ghost: The Origin of Species Updated*. New York; Random House, 2000.

Killick, Rachael. "Family Likeness in Flaubert and Maupassant: 'La Légende de Saint Julien l'Hospitalier' and 'Le Donneur d'eau bénite'." *Forum for Modern Language Studies* 24.4 (October 1988): 346–358.

Lecarme, Jacques. "Sartre lecteur de Maupassant." *Lectures de Sartre*. Ed. Claude Burgelin. Lyon: Presse Universitaire de Lyon, 1986. 185–198.

Ledwon, Lenora. "Darwin's Ghosts: The Influence of Darwinism on the Nineteenth-Century Ghost Story." *Proteus* 6.2 (Fall 1989): 10–16.

Lefebvre, René. "Faire son propre malheur: Maupassant romancier face au retour du Même" *Littératures* 38 (Spring 1998): 79–100.

———. "Horreur et ressemblance dans les contes et nouvelles de Maupassant: Du semblable à l'innommable." *Travaux de Littérature* 13 (2000): 227–245.

Lintvelt, Jaap. "L'Homme et l'animal dans les Contes de Guy de Maupassant." *Acta Litteraria Academiae Scientarium Hungaricae* 32.1–2 (1990): 71–78.

———. "L'Animalisation humaine et l'amour dans 'Berthe' de Maupassant." *L'Homme et l'animal*. Ed. Sandro Briosi and Jaap Lintvelt. Groningen: Rijksuniv. Groningen, 1988. 65–71.

Maleuvre, Didier. "Can We Believe Darwin?" *Comparative Literature* 53.2 (Spring 2001): 117–130.

Marks, Jonathan. *What it Means To Be 98% Chimpanzee: Apes, People and Their Genes*. Berkeley: University of California Press, 2002.

Mayr, Ernst. *Toward a New Philosophy of Biology: Observations of an Evolutionist*. Cambridge, MA: Harvard University Press, 1989.

Mead, Gerald. "Social Commentary and Sexuality in Maupassant's 'La Maison Tellier'." *Nineteenth-Century French Studies* 24.1–2 (Fall 1995-Winter 1996): 162–169.

Millstone, Amy B. "Maupassant's 'Bed 29,' or the Case of the Tainted Patriot." *Romance Languages Annual* 8 (1996): 59–65.

Murphy, Ann L. "Narration, the Maternal, and the Trap in Maupassant's Mademoiselle Perle." *French Review* 69.2 (December 1995): 207–216.

Orlich, Ileana Alexandra. "Tracking the Missing Link: Maupassant's 'Promenade' and James's The Beast in the Jungle. *The Comparatist: Journal of the Southern Comparative Literature Association* 18 (May 1994): 71–89.

Place-Verghnes, Floriane. "Maupassant pornographe." *Neophilologus* 85.4 (October 2001): 501–517.

Prévost, Maxime. "Le Cadavre de Schopenhauer dans le cercueil d'Edgar Poe: Notes sur Auprès d'un mort de Guy de Maupassant" *Neophilologus* 84.3 (July 2000): 371–383.

Sillam, Maguy. "Bel-Ami, héros de la transgression." *Cahiers Naturalistes* 42.70 (1996): 83–92.

Smith, Diane M. "The Evolution of the Working Class Novel in Europe: Darwinian Science and Literary Naturalism." *Excavatio* 8 (1996): 72–85.

Soubias, Pierre. "La Place de l'Afrique dans l'imaginaire de Maupassant: Une lecture des nouvelles africaines." *Maupassant multiple*. Ed. Yves Reboul. Toulouse: Presse Universitaire du Mirail, 1995. 29–39.

Suwala, Halina. "Maupassant 'salonnier'." *Excavatio* 15.3–4 (2001): 198–208.

Turcanu, Radu. "Le Horla et 'l'effet de réel'." *Nineteenth-Century French Studies* 26.3–4 (Spring-Summer 1998): 387–397.

Viti, Robert M. "The Elemental Maupassant: The Universe of *Pierre et Jean*." *French Review* 62.3 (February 1989): 445–455.

Weiner, Jonathan. *The Beak of the Finch*. New York: Vintage, 1995.

———. *Planet Earth*. Toronto, New York, London, Sydney and Auckland: Bantam, 1986.

Willi, Kurt. "Déterminisme et liberté chez Guy de Maupassant." Zurich: Juris-Verlag, 1972.

Zouaghi-Keime, Marie-Anne. "Violence et démence dans l'oeuvre de Maupassant." *Violence et fiction jusqu'à la Révolution*. Ed. Martine Debaisieux and Gabrielle Verdier. Tübingen: Narr, 1998. 99–108.

Index

A

B

C

D

S

T

U

V

Y

Z

Currents in Comparative Romance Languages and Literatures

This series was founded in 1987, and actively solicits book-length manuscripts (approximately 200–400 pages) that treat aspects of Romance languages and literatures. Originally established for works dealing with two or more Romance literatures, the series has broadened its horizons and now includes studies on themes within a single literature or between different literatures, civilizations, art, music, film and social movements, as well as comparative linguistics. Studies on individual writers with an influence on other literatures/civilizations are also welcome. We entertain a variety of approaches and formats, provided the scholarship and methodology are appropriate.

For additional information about the series or for the submission of manuscripts, please contact:

Tamara Alvarez-Detrell and Michael G. Paulson
c/o Dr. Heidi Burns
Peter Lang Publishing, Inc.
P.O. Box 1246
Bel Air, MD 21014-1246

To order other books in this series, please contact our Customer Service Department:

800-770-LANG (within the U.S.)
212-647-7706 (outside the U.S.)
212-647-7707 FAX

or browse online by series at:

www.peterlangusa.com